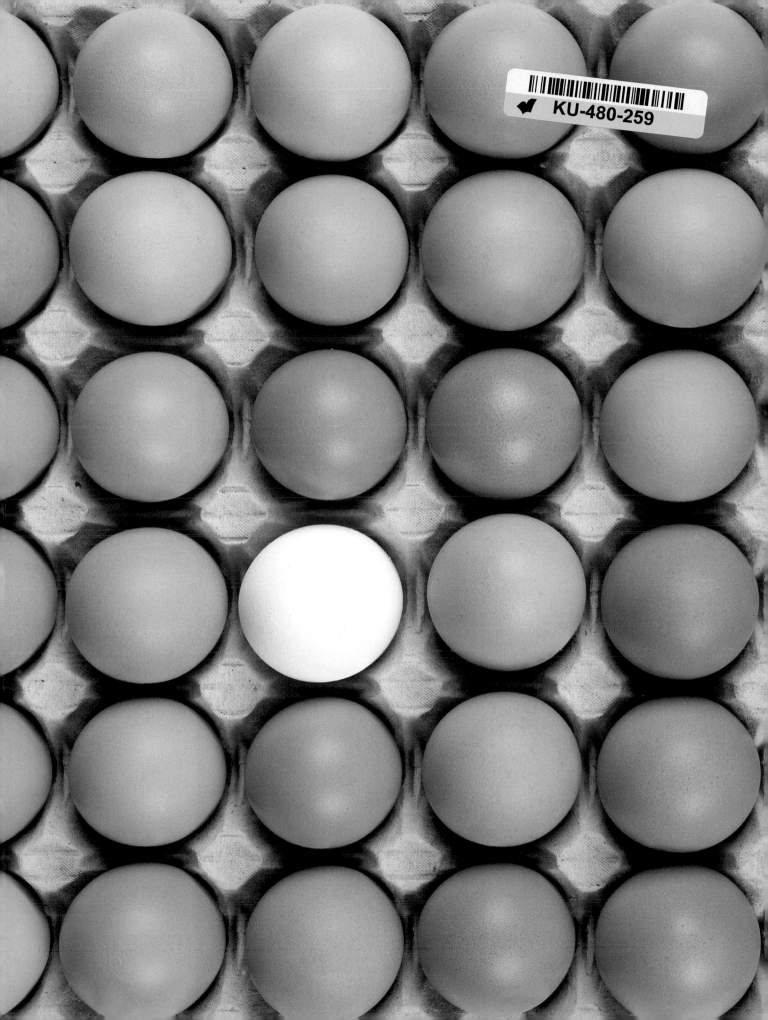

Haynes

Chicken
Manual

First published in February 2010
Reprinted in 2010 (twice) and 2011 (twice)

British Library Cataloguing in Publication Data
A catalogue record for this book is available from
the British Library

ISBN 978 1 84425 729 4

Published by Haynes Publishing,
Sparkford, Yeovil, Somerset BA22 7JJ, UK
Tel: 01963 442030 Fax: 01963 440001
Int. tel: +44 1963 442030 Int. fax: +44 1963 440001
E-mail: sales@haynes.co.uk
Website: www.haynes.co.uk

Haynes North America Inc.
861 Lawrence Drive, Newbury Park,
California 91320, USA

Printed in Italy by G. Canale & C. S.p.A.

**While every effort is taken to ensure the accuracy of
the information given in this book, no liability can be
accepted by the author or publishers for any loss,
damage or injury caused by errors in, or omissions
from the information given.**

Credits

Author:	**Laurence Beeken**
Project Manager:	**Louise McIntyre**
Copy editor:	**Ian Heath**
Page design:	**Richard Parsons**
Index:	**Alan Thatcher**
Illustrations:	**John Lawson**
Photography:	**Laurence Beeken**
	istockphoto.com

Haynes

Chicken
Manual

The complete step-by-step
guide to keeping chickens

Haynes
®

Laurence Beeken

CONTENTS

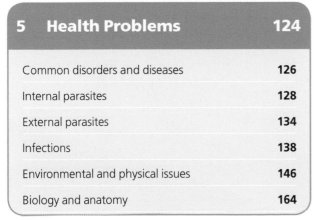

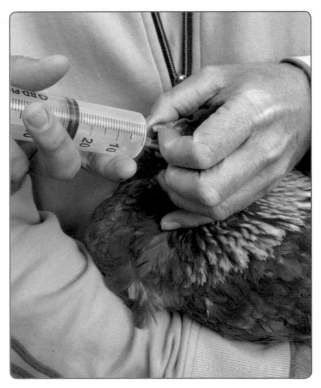

INTRODUCTION

Why keep chickens?

Up until as recently as the 1950s it was the norm for many households – even in inner city areas – to keep poultry alongside their vegetables. It was a practical way to get fresh eggs and the occasional roast bird, and a means of controlling pest infestations in the garden, so that in effect chickens contributed as much to the kitchen garden as plants did. Unfortunately the rise of post-war consumerism and the ensuing availability and cheapness of meat from the new 'super' markets more or less saw an end to this aspect of domestic gardening. Regular salmonella scares in the 1980s and, more recently, concerns over bird 'flu (Avian influenza) have since resulted in further decline, and now most gardeners seem to have largely forgotten that chickens once played an integral part in sustaining the family.

Fortunately, however, there has recently been a resurgence of interest in keeping backyard chickens, due in no small part to concerns about welfare, pesticides and genetically modified foods, as highlighted by the proliferation of lifestyle programmes on our television screens as we sit down to our evening meals.

For the kitchen gardener there are several very good reasons to employ a few birds around the place, as chickens are great company, especially for children; and as far as your household is concerned they'll earn their keep by providing enough eggs for a small family to cook with and eat. Furthermore, since you control their intake of food and water you'll know that your eggs are from happy healthy chickens, not a factory farm, and need have no fears of them being affected by antibiotic treatments or anything genetically modified.

Above: Children are fascinated by poultry.

An additional benefit is that in return for a balanced layers' ration and the ability to free-range, your hens can be counted on to seek out and destroy all manner of pests that might infest your garden and lawn, and will produce high-quality manure which, when combined with your compost, will make an ideal fertiliser for your plants.

Where to put them

How many times have you thought 'If only I could do something with that bit of garden,' or 'It's about time that weed bed was sorted out'? Well, that piece of spare ground

Below: Chickens are an essential part of the kitchen garden.

Below: Free range is the best option for your birds.

might be just the area where a few chickens could thrive, if properly managed. Keeping chickens there will make it productive and will create an area that suddenly the whole family is interested in, particularly the children, who seem to delight in the funny ways of poultry – for them it will also be an educational experience, teaching them about livestock and providing an understanding of the practical issues relating to husbandry.

You are what you eat

Since you have full control over the diet that your home-kept birds are fed, both in quality and variety, you can contribute directly towards the nutritional content, so that eggs from your own hens may well be healthier for you. There is some evidence that free-range eggs contain a greater amount of folic acid, Vitamin B12 and Vitamin A than battery eggs, and with a more varied diet the eggs produced by your own chickens may well be even better for you than those from commercial free-range birds. Certain types of egg, such as those from Araucanas, may even be lower in cholesterol, and since eggs in general contain less harmful cholesterol than was originally thought, if you remove the yolk you have a high protein, low fat snack. On the downside, there are increasing consumer concerns regarding the levels of potentially dangerous drug, hormone and antibiotic residues that may be present in intensively produced eggs.

Home-produced eggs

Today, eggs purchased from supermarkets and shops are generally of high quality and freshness thanks to initiatives such as the Lion Quality stamp, which means that people no longer have to break an egg into a cup first to check that it hasn't gone off. Despite this, however, most backyard poultry-keepers will still tell you that their eggs are far superior to shop-bought eggs of whatever type – and it's certainly true that if you've never cracked a home-produced egg into a frying pan and watched the deeply coloured yolks sizzle, then you really don't know what you are missing.

Just as people say that home-grown vegetables are much tastier, the same could be said when comparing home-produced eggs to those supplied through a supermarket. You only have to crack one brought from a shop to notice the thinner shell and the watery contents, whereas your own carefully fed free-ranging birds will lay eggs with a fantastic yellow yolk and a thick white – not a bit like some insipid shop-bought ones which have often been artificially enhanced with colorants to make them more appealing to unsuspecting shoppers.

Above: The lion stamp guarantees quality.

Table birds

Whilst some keepers may want eggs, and some want pets, others find it more useful to keep poultry for meat, and their chickens are consequently referred to as 'table birds'. Understandably, the thought of killing the birds may be abhorrent to some, but just like home-produced eggs, those who have made the leap to home-reared table birds will never go back to shop-bought poultry. But if it is your intention to keep table birds then it must be clearly understood and accepted from the start that the day will come when they must be killed, dressed, and eaten, and

Below: Light Sussex make an ideal table bird.

Above: Free range gives a much better texture to the flesh of table birds.

unless you're prepared to do this humanely and properly then this is not the option for you.

As far as the quality of meat is concerned, a bird kept free-range will have a more meaty texture to its flesh as well as a better flavour, as it has been able to range freely, allowing the muscles to become developed and rich in blood supply and therefore nutrients, meaning in turn that its bones are stronger and less likely to splinter. Quite often birds sold as free-range in the shops have only met the minimum requirements laid down to qualify as such, and if you compared their meat to your own birds you'd be very much surprised by the inferior taste.

A breed that attains a good size for the table and has hens which produce a good quantity of quality eggs in the laying year is referred to as a utility breed, and it is from the original utility birds that all of today's breeds and hybrids have descended. A breed that produces birds of both table and laying quality is described as dual purpose.

Below: Showing is a rewarding hobby.

As a hobby

Living in a world with so many stresses and strains, keeping poultry can be more than just a productive hobby. One of the most regularly prescribed cures for people who have stress-related illnesses or have suffered heart problems is to get a hobby, while someone nearing retirement age is frequently advised to 'keep yourself occupied'. And as a hobby, keeping poultry has one distinct advantage over, for example, fishing – which is that it's a seven days a week, 52 weeks a year pastime; in other words, it generates continued interest.

Depending on the time and money available to you, keeping chickens for a few eggs can eventually progress to an interest in exhibiting purebreds, which was one of the most popular and high-profile pastimes of the Victorian era and is today slowly re-emerging as a popular hobby. Many new and experienced owners, having found hobby keeping to be fun, are only too willing to show off their birds, perhaps starting with a local summer show and then progressing, with experience, to the larger national events, where winning brings attention to you as a breeder and attracts a higher price for your stock.

If you find exhibiting live birds too much trouble but enjoy the competitive spirit, there are always egg classes too where competition is just as buoyant and equally fierce.

Welfare concerns

Nowadays, more than ever, we are increasingly aware of animal welfare and organic culture, so one compelling reason for keeping your own chickens must be that their lives will be better and you need not feel the guilt of suspecting that your food might be the product of cruelty or neglect. Even when you're rearing a bird for meat, you're safe in the knowledge that it lived a comfortable, healthy life before being quickly and painlessly dispatched without being subjected to the trauma of travel.

Below: Battery hens are a sorrowful sight when first rescued.

Above: The Sultan is a rare breed worthy of conservation.

Conservation

The Rare Breeds Survival Trust (www.rbst.org.uk) is the leading conservation charity working to restore Britain's native livestock breeds to their rightful place in our countryside, and is funded entirely by membership subscriptions, donations and legacies. As with many things fads and fashions come and go, and it is no different with poultry, where different breeds become popular for a season and are then forgotten. The hobby keeper can do much to help by keeping a few rare breeds as part of the garden flock (as long as you remember to keep them segregated from other breed males during mating), and thus maintaining a genetic pool for the future.

Selling and profit

You'll never become a millionaire by selling chickens for profit – in fact most people will only tend to break even, especially as feed costs continue to rise. But having said that, chicken-keeping is still a hobby that can pay for itself when you sell birds, eggs and meat (taking into account any relevant legislation, for which you should visit the Department of the Environment, Food and Rural Affairs website at www.DEFRA.gov.uk/foodrin/poultry/trade/index.htm).

Eggs can be sold on a small scale to neighbours and friends, and you'll find a steady stream of people only too willing to take a slaughtered table bird off of your hands. Fertile eggs can be sold via eBay as well as the sales sections of certain poultry websites, although there are website and legal guidelines that you'll need to research. Good exhibition results will certainly help you achieve higher prices, as will a keen eye for the next popular breed or colour.

Local farmers' markets and established poultry auctions are ideal for selling live birds, but again you need to be fully conversant with the pertinent regulations – both statutory and bespoke to the market – before selling, as a bad bird will be associated with your name well into the future.

The selling of sundries and equipment may well be a more profitable option if you're of a mind to set up your own business, but a thorough knowledge of poultry is essential if you're to compete with the major players in the market.

The law

Under new laws introduced to control Avian influenza, you must register with the Department of the Environment, Food and Rural Affairs (DEFRA) if you own or are responsible for a poultry premises with 50 or more birds.

This requirement also applies even if the premises are only stocked with 50 or more birds for part of the year. At present premises with fewer than 50 birds are not required to register, based on expert advice that smaller flocks are less likely to play a significant role in the spread of disease, but keepers are still encouraged to register voluntarily. In addition to this, if you're a poultry breeder with over 100 birds you'll also need to register your premises and apply for a registration number, which will need to be quoted on all dispatches from your premises. Full details can be found at www.DEFRA.gov.uk/foodrin/poultry/trade/marketregs/hatchingregs.htm (click the link that relates to the production and marketing of hatching eggs).

The regulations don't necessarily mean or refer to 50 of one species (although this may be the case) but 50 overall – for example, if you have 30 chickens, 15 geese, 4 ducks and a rhea you'll still need to register.

Below: Eggs can still be sold at the gate.

History of the chicken

Above: Many current Asiatic Fowl resemble the original Red Jungle Fowl.

Commonly referred to as fowl (wild state) or chickens (domesticated state), the origins of the domestic chicken remain the subject of much debate between archaeologists and geneticists alike, although most will agree that the red jungle fowl (scientific name *Gallus gallus*) is the main ancestor of the domestic species, while DNA information obtained in the 1990s suggests that one population of red jungle fowl – most likely the South-East Asian Red Jungle Fowl – is likely to have provided the maternal ancestor of all domestic chickens.

Below: Marco Polo identified the Silkie during his travels.

Genetically, the genus *Gallus* is composed of four species: *G. gallus* (Red Jungle Fowl), *G. lafayettei* (Lafayette's Jungle Fowl), *G. varius* (Green Jungle Fowl), and *G. sonneratii* (Grey Jungle Fowl), along with five sub-species of Red Jungle Fowl, *G. g. gallus* (South-East Asian Red Jungle Fowl), *G. g. spadiceus* (Burmese RJF), *G. g. bankiva* (Javan RJF), *G. g. murghi* (Indian RJF) and *G. g. jabouillei* (Vietnamese RJF). These classifications are mainly based on physical traits and geographic distribution of the populations and it is likely that all have contributed in some part to the make-up of the present domestic chicken, which can be considered to be either a sub-species of Red Jungle Fowl (*G. g. domesticus*) or a separate species (*G. domesticus*).

The Red Jungle Fowl found in the forests of South-East Asia and India would have spread during the process of domestication as nomadic peoples travelled to other parts of the world by land and sea. Archaeological evidence suggests that chickens moved from the Indonesian islands to mainland Asia via Thailand and then onwards through southern Asia into India and China, before ending up in Europe and the Americas. Subsequent breeding programmes resulted in time in breeds of chicken representing four distinct lineages: egg-type, game, meat-type and bantam.

History is full of references to domestic chickens, from Polynesian seafarers who reached Easter Island in the 12th century AD along with their poultry (which they housed in extremely solid chicken coops built from stone), to Marco Polo, who wrote of hens which had 'hair like cats, were black and laid the best of eggs' (he was almost certainly referring to Silkies, which have barbless feathers that look and feel like fur).

Cockfighting

Male chickens – known as cocks, cockerels (if younger than one year old), or roosters – are common symbols of masculinity, and their natural inclination to fight has in the past been exploited in cockfighting contests, sometimes with a metal spike added to or in place of their natural spurs. The combatants, known as gamecocks, were (and in parts of Asia still are) specially bred birds, conditioned for increased stamina and strength. Their combs and wattles were traditionally cut off in order to limit blood loss when damaged during fighting and to prevent freezing in colder climates, and this custom was subsequently incorporated into the show standards of both the American Gamefowl Society and the Old English Game Club, where 'dubbing' the comb and wattles (as it is known) became normal practice.

Possessing an inherent aggression toward all males of the same species, cocks were given the best of care until

nearly two years old, being conditioned much like professional athletes so as to reach their peak prior to events or shows. Wagers were often made on the outcome of the match, with the unfortunate combatants either fighting to the death or enduring such physical trauma that it resulted in death.

Although most countries have long since banned cockfighting (with the exception of parts of South-East Asia), it was a popular sport in ancient Greece, where pictures of chickens have been found on Greek red-figured and black-figured pottery and, according to tradition, fighting cocks were fed with garlic and onions to increase their aggression.

Chickens in religion

Although ritual sacrifices were at the heart of Greek worship, chickens were not normally used for this purpose since at that time they were still considered to be exotic animals, and were considered attributes of Ares, Heracles and Athena. The Greeks believed that even lions were afraid of cocks, and such was the associated power of the cockerel that the gift of a fighting cock among men was seen as a common way to initiate a homosexual relationship.

The Romans used chickens as oracles, where the hen gave a favourable omen if she appeared from the left. Although any bird could be used only chickens were normally consulted, and were cared for by a special keeper called the *pullarius*, whose job it was to open the chickens' cage and feed them when an augury was required. It was deemed a bad omen if the chickens remained in the cage, clucked, flapped their wings or flew away; but if they ate hungrily the omen was considered to be good. In 249 BC the Roman consul Publius Claudius Pulcher had his chickens thrown overboard when they refused to feed – in effect a bad omen – before the naval battle of Drepana, 'so that they might drink since they refused to eat'. He promptly lost the battle against the Carthaginians, 93 Roman ships were sunk, and he was subsequently tried and found guilty of incompetence and impiety and was heavily fined.

In many Central European folk cultures the Devil was believed to flee at first cock crow, while in Bali the blood shed after a cockfight was regarded as cleansing.

Right: Even today the chicken is a potent religious symbol.

The golden age of chicken-keeping

Although poultry keeping dates back many centuries, it was only in Victorian times that a Standard of Type was written down for specific breeds. The abolition of cockfighting in England and Wales as part of the Cruelty to Animals Act in 1835 had resulted in breeders having nowhere to show off their stock, and as good stock promoted good prices an alternative showcase was urgently needed, and poultry exhibitions came about partly to fulfil this need. Birds were judged on physical attributes based on what their particular breed was created for, so a laying bird would have a particular width of back, a meat bird a specific depth of breast and a game bird a particular stance – and so the breed standards were set.

Above: The Golden Era of poultry.

Below: Consumer pressure is reducing the amount of battery farms.

The presentation by Sir Edward Belcher to Queen Victoria just a few years later, in 1843, of five young hens and two cockerels from a breed of Vietnamese Shanghai Fowl (now known as Cochins) caused quite a stir in social circles. In 1846 birds from the Royal Aviaries were shown at the Royal Dublin Society show, where they were awarded the Gold Medal – and so began the golden era of poultry exhibitions, causing both competition and prices to rocket. Even today the Poultry Club of Great Britain has Prince Charles as a patron, with Buff Orpingtons being amongst his flock.

Having Royal patronage gave pure breeds and poultry-keeping in general an enormous boost, and poultry mania quickly caught on to such an extent that when one of the original poultry importers held an auction, his 150 lots sold for £606 – which, when you consider that the average wage at the time was probably around £60 per annum, gives you some idea of how lucrative the trade was. Today respected lines can still fetch as much as £100 per bird, although this is in no way comparable to the astronomical level of Victorian prices.

Poultry-keeping was revolutionised after World War Two. In 1945 less than 1% of laying hens were caged, but this percentage rose rapidly thereafter until, by 1986, nearly 93% cent of the national flock were kept in cages.

Today

Battery cages are now being phased out in the UK, as more consumers have become aware of, and opposed to, the welfare implications; and with the surplus hens being re-homed by welfare organisations free-range eggs are becoming ever more popular, so that nearly one in four British hens are now free-range.

Intensification of chicken-breeding for meat production

Above: Exhibitions now are much smaller than during their heyday.

has become equally extreme, as pre-war it took 126 days to produce a 4lb (2kg) bird while now it takes only 42 days. But there are increasing welfare concerns about broiler chickens (birds bred specifically for the meat trade), and many people prefer to buy poultry that has been reared in less intensive conditions, whereby the meat is of better quality, has a more pronounced taste, and comes from birds that have led more enjoyable lives with fewer health problems.

Poultry exhibiting has also undergone significant change. As with most animals exhibited as a hobby, the birds shown today bear little resemblance to the cockfighters and exhibition specimens of yesteryear, having in many cases lost the very attributes that they were originally bred for in favour of fluff and visual appeal. Many a good exhibition line is no longer capable of laying more than 40 or so eggs a year (although there are breeders working on improving the laying capacity of older traditional breeds, such as Dorkings), and pure lines have been replaced by hybrids created to produce upwards of 280 eggs per year.

Outbreaks of bird flu and salmonella scares have done much to decimate the showing circuit, with many breeders cutting back on stock to devote their efforts to a few valuable bloodlines while others no longer attend shows for fear of their stock becoming infected. Showing has consequently dipped to an all-time low, although the recent revival of interest in poultry-keeping means that it is slowly getting back on its feet. However, it's unlikely that it will ever recover to the level of popularity it enjoyed during the Victorian era.

Although the future of exhibitions depends very much on the attitude of governments, and on public perception of the hazards as portrayed in the media, it should be remembered that show birds were created first and foremost as a sporting attraction, and there will always be a core of enthusiasts who want to show their hard work through the medium of exhibitions.

Above: More and more chickens are kept free-range.

Meanwhile increasing interest in organically reared food, and its promotion by celebrities, is resulting in even greater awareness of the advantages of keeping your own chickens, whether as family pets (the prerogative of a civilised society) or as an addition to the kitchen garden.

Chickens, poultry, chooks – whatever you wish to call them, they're here to stay.

Below: Chickens are here to stay.

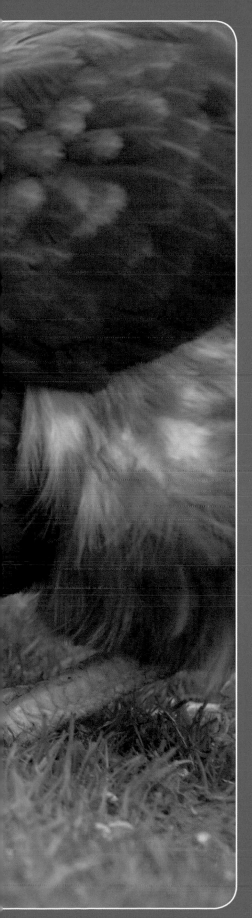

GETTING STARTED

Preliminary considerations

Ask yourself what you need in order to lead a comfortable existence: a home, security, food and water, a job, companionship, and care when you're unwell. In all of these respects, chickens are no different.

The coop or house you eventually decide on for your chickens must be free from water ingress and draughts, and provide security from predators and danger (meaning it must be able to prevent them from roaming all day to annoy your neighbours and cause traffic chaos, and at night will ensure that foxes can't get to them). It may therefore also be worth considering an integral run, or installing an electric fence around them – which, contrary to popular belief, is neither cruel nor dangerous as it provides no more than a brief, harmless shock that will deter predators and will keep your chickens where you want them (see page 82). There are many Internet-based suppliers when it comes to solutions in these respects.

Of course, you'll need birds to put in the house, whether rescued battery hens, commercial hybrids or pure breeds, and you'll need to think carefully about what you want to get out of the relationship that you'll foster, be it as pets, for eggs, for meat or for conservation and showing. You should also bear in mind that you don't need a cockerel in order to make the hens lay.

A good feeder and drinker are a necessity. They need to be movable in case you need to get the birds under cover,

and to make them more practical when it comes to cleaning them out. With drinkers, a traditional green- or red-bottomed one is the best for a few birds, and a larger automatic one for a larger group of, say, six or more. Make sure that when the birds drink they can get their heads into the base, as this is quite a common cause of dehydration in older birds with larger head ornamentation, especially when introducing them to their new quarters after you've acquired them. Fill the containers with fresh water daily and supply a good-quality mash or pellets in the feeder. Don't be tempted to scrimp on cost here, as you'll only get out what you put in.

Feeders can be of the galvanised type, or plastic, which will invariably have a red bottom to attract the birds (chickens are drawn to red, which makes wound pecking a problem). You'll also need a bowl for flint grit, or more practically a flowerpot with a wire threaded through the drainage holes in the bottom.

Which feed you ultimately use is a matter of preference, what the birds are used to, and a bit of trial and error. Both layers' pellets and mash are compound feeds specially formulated to provide all the essential nutrients and vitamins needed. If you have growers or young stock, then you can fill a hopper and allow them access as they want, just making sure that wild birds can't foul the container or eat from it, as it's too costly to feed all the local birds your precious pellets and they may introduce disease.

Below: Do not allow your chickens to wander.

Below: A few birds will drink happily from a green bottomed drinker.

Above: Access to vegetation is important.

Access to vegetation is equally important, whether in a run or hung up in the pen, as the birds will digest cellulose in a separate part of the gut in order to absorb the nutritional element. There's also the physiological issue to consider, as chickens are in origin forest birds and vegetation is important to their emotional wellbeing.

At some point you'll need to clean the house out. Try not to be too finicky here, as a pristine house – while nice to look at – doesn't allow the birds to build up a basic immunity to the low-level challenge offered by a bit of muck. You should invest in a rubber trug, a coal shovel, possibly a wallpaper scraper, and either good-quality wood shavings or paper shreddings (cross-shredded, not the long ones) for fresh bedding. A bale of straw will never go amiss, and if nothing else provides endless entertainment and shelter for the hens.

These, then, are the basics: a coop, a run or fence, and feed and water supplied in proper dispensers.

Below: The correct cleaning tools are essential.

Consider the issues

Before embarking on your chicken-keeping experience, it's essential to ask yourself some basic questions and consider the pros and cons, which will establish if indeed chickens are for you and, if so, where you're going with them?

The pros
- Supply of fresh eggs.
- No restrictions on keeping a few birds.
- Relatively cheap to acquire.
- Relatively cheap to maintain.
- Hens are generally quiet.
- Surplus eggs can be sold.
- A hobby that can pay for itself.
- Table birds can be reared.
- Entertaining and educational for children.
- Good family pets.
- Garden pest clearance.
- Source of fertiliser.
- Conservation/rescue opportunities.

Cons
- DEFRA require registration of a flock of 50 or more.
- Can destroy flowerbeds and lawns if allowed to roam.
- Cockerels are noisy.
- Clauses in deeds may forbid keeping poultry.
- Pure breeds may not lay well.
- Availability of affordable housing and feed.
- Some pure breed lines are expensive.
- Vulnerable, and constant security is needed.
- Local vets may not understand poultry.
- Need daily care – can't be left if you go on holiday.
- Poor husbandry attracts vermin.
- Escapees may get into neighbours' gardens or cause accidents.

Below: A crowing cock can be noisy.

That said, any of the above points will have an argument to the contrary. For example, you may get lots of eggs, but what happens if you can't sell the surplus? Leftover feed attracts vermin, but if you're clean in your daily chores then there'll be nothing for them to eat and they won't become a nuisance. Lawns may be scratched bare in the winter, but the added fertiliser will make them grow with added vigour in the spring. So, your decision should be as educated as possible based on the information you have to hand.

If you're able, it may be worth asking a friend or neighbour who keeps chickens if you can spend an hour or so with them to see what exactly is required. Summer shows and sales are an excellent opportunity to gain more information, and you should also read as much as you can on the subject. Whatever you do, don't approach poultry-keeping unprepared – it will only lead to disappointment and failure.

Chickens in the garden

Perhaps you'd prefer to keep your birds as a part of your garden, and aren't bothered about the odd piece of mayhem here and there. Chickens do provide an excellent pest-clearing service while at the time keeping the soil surface tilthed and removing many weed seeds. In addition, their waste provides superb fertiliser, which must be rotted slightly before use as fresh manure will burn delicate foliage.

Many kitchen gardeners and allotment holders use their birds as part of their rotation system, housing them directly on the beds in the autumn to seek out and eat insect larvae and fertilise the area ready for spring plantings. Hybrids are an

Below: Chickens provide an excellent pest clearance service.

Wing clipping

Clipping one of a chicken's wings is carried out to unsteady the bird in order to prevent it from attempting to fly over a barrier. You should clip the wing annually, to ensure that the old clipped feathers haven't moulted out and been replaced by new ones. And you should only ever clip one wing – if both wings are clipped, you won't actually unsteady your chicken, only hamper it until it eventually learns to compensate.

Take one wing in your hand (it helps to have another person holding the bird when you do this for the first time) and, using a sharp pair of scissors, cut halfway down and along the first ten feathers (the primary feathers), through the quill. If you cut nine or eleven, don't worry – the figure of ten is a guideline only, as this is how many primary feathers there are.

It's a painless process for the bird, similar to us cutting our nails, although you shouldn't attempt to cut immature feathers just after the moult, as they may still contain blood and will bleed.

excellent choice for this purpose, and probably a lot easier for the novice to keep, as they tend to be very hardy and have been bred specifically for meat and/or egg production (ie utility), so that you get a return for letting them wander.

If you want to do your bit for conservation you can't get much better than a good utility strain of Light Sussex, although you should try to avoid exhibition stock, as these tend to have lost their laying capacity in favour of looks. Feather-footed breeds such as Cochins, Pekins and Sultans tend to get too muddy if left in the vegetable garden and soon become dishevelled and open to stress-related illnesses.

Most people will opt to keep a few hens in the flower or vegetable garden, but this is not without its daily challenges. In order to keep your chickens happy and yourself stress-free, a few simple guidelines can be followed to great effect:

Keep everything out of reach

An obstruction such as a decorative fence panel or a low hedge, such as box (*Buxus sempervirens*), is ideal, and is particularly good if you're segregating the kitchen garden, as the design lends itself to formal lines. Birds are quite lazy in

Above: Keep plants in containers to avoid damage

their meanderings, and smaller, chunky breeds such as Pekins, with their foot feathering, find it too much effort to jump, and a simple clip along the flight feathers of one wing will unbalance them sufficiently to prevent flying (see box). The larger, heavier breeds such as Cochins and Brahmas will also be deterred by a barrier, while the ragged feathers of Silkies stop them even attempting to cross it by keeping them firmly on the ground.

Fencing is suitable if you have the space and money and you want to limit access to an area completely, but many owners want to enjoy their birds, and part of the attraction of chickens is the amusing antics that they get up to, which will keep you amused for hours.

Containers are a contemporary and much under-utilised solution to the problem of preventing access, and in this age of ever smaller gardens and minimalist planting there are many alternative options. The traditional hanging basket can be easily used for strawberries, peas or basket tomatoes, or anything really that hangs or trails, as the advantage here is that these can be hung out of the way of most poultry, with the exception perhaps of some of the smaller more agile breeds, such as Dutch Bantams. Window boxes can accommodate lettuces and herbs, while larger pots can contain all manner of perennials and vegetables; and even if the most determined of birds has the insight to jump into the pot, there's very little damage that they can do in the limited space available.

Choose your plants

Taller plants and shrubs do tend to fare better; use roses, berberis, dogwoods and hydrangeas, as all are both colourful and versatile in a mixed border while the first

two come with their own spiky defences. Raspberries are excellent for planting in a chicken run, as the lower berries provide fun for the birds while the higher ones can be saved for the family.

Fast-growing, woody stemmed annuals can be pot grown and then planted out when they are a size suitable to withstand onslaught. Tall fleshy stemmed perennials should be staked where necessary, as they can be damaged by the chickens' wanderings, and while the breed may not necessarily fly they can and do jump.

If you want to grow flowers why not go for edible ones? Sunflowers are ideal, as their sturdy stems can survive the worst of a chicken's attentions and they grow high enough for the flowers to escape beaks. When you've enjoyed the colourful flowers you can dry the seed heads for your poultry.

Choose your breed

Smaller breeds do tend to be less punishing in their scratchings, but their size makes it easy for them to fly over obstructions to reach their goal. Larger breeds tend to respect barriers, but can cause mass destruction when their feet make contact with any vegetation. Consider feathering too. Silkies with their barbless feathers find it difficult to fly too high, while crested breeds such as Polands sometimes have difficulty seeing their target and aren't really suitable if you can't provide shelter.

Whatever you do, there will be both triumphs and failures. One bird will develop a penchant for peas, while another will ravage the radishes. There is no accounting for taste, as chickens are intelligent animals and have their own individual traits. Let your preference for style guide you, and let your birds decide the outcome, but above all have fun.

Below: The full crest of a Poland makes it difficult for them to see their target

Choosing a breed

Above: The Dutch is a True Bantam.

Chickens come in various sizes – large, bantam (or miniature), and true bantam – so the size of the bird that you ultimately choose will reflect on the space that you have available and the degree of activity that you'll find desirable. True bantams are those small breeds that have no large equivalent, such as Japanese and Dutch, while bantams or miniatures are approximately one-quarter the size of their large fowl counterparts, for example bantam Araucanas and Wyandottes.

Think about what appeals to you and the purpose you want to keep them for. If you want eggs, then it's probably best to start with a hybrid such as a speckled hen or a brown layer, or perhaps a utility pure breed which will have been selectively bred for egg-laying capacity, such as Sussex or Marans.

Below: Hybrids will give you a good supply of fresh eggs.

You may not want one of the breeds that go broody at the drop of a hat, such as the Silkie (although they'll come back into lay towards the end of winter, making them invaluable when most other breeds are refusing to lay at all). Consider if you want a table bird, in which case select either a commercial broiler or, again, a pure breed specially selected for table quality, such as La Flesche. Decide if you want a breed for exhibition or as a pet, as the characteristics bred into each type will vary – for example, Pekins are very docile, while the active Modern Game will give you a run for your money.

Some breeds are aggressive or flighty, with birds taking to the air at the least provocation, some will be noisy, and others will be fairly quiet (although all breeds will make a noise to differing degrees). Think about the space you have available, as size and temperament could mean that you're trying endlessly to get a stubborn Leghorn out of a tree, or being chased by a particularly aggressive Asil.

Remember that larger birds will produce more manure, so if you don't want to be endlessly cleaning housing pick a smaller breed. If the time you can spare to spend with your birds is limited then exhibition stock is probably not for you, and you should also avoid breeds with specialist features such as beards, crests and feathered feet, since after a few days outside these poor birds will look extremely bedraggled and become stressed, resulting in increased susceptibility to disease and more upset for you.

Having said that, though, feathered-footed birds are less enthusiastic diggers, so if you want to protect your flower beds then something like a Pekin or a booted bantam may be for you, as long as you provide a dry perch at night so that the foot feathering can dry.

Below: Certain breeds are natural exhibitionists.

Certain short-legged bantams such as the Japanese will have problems getting around, and frizzled-feathered birds don't do well in the rain, so how you apply shelter and space to roam will also impact on your decision. How much time you can spend on research will have a bearing on your ultimate choice, since if you just learn the basics a hybrid such as the Black Rock or a tough breed such as the Marans will be your best option, leaving something like the Derbyshire Redcap and its delicate comb until you have more experience.

There are breeds aplenty in all manner of sizes, shapes and features, and all will have different characteristics that will appeal to you. There are breeds that love to fly, those that are great escapologists, and others that are very docile. Be warned too that your chickens will take on your own characteristics, so if you're of a nervous disposition your poultry will become agitated when you go near them, in which case you'd be better off again with a more docile breed.

As a final point, select your breeder carefully too, since often they'll have bred certain traits into their own flocks, either consciously or in error, so you may find a particular breeder has fluffier Orpingtons than others or more aggressive Shamos. For more information on the wide variety of pure breeds available refer to the book *British Poultry Standards*, written by Victoria Roberts in association with the Poultry Club of Great Britain.

The following breeds are some of the more popular of the vast array available, and while each variety will have similar requirements, they'll exhibit one or more traits specific to breed, sex age and even colour.

Above: Hybrids are an excellent choice for the novice.

Below: Chicks are imprinted within three days of hatching.

Fact...

Chicks are imprinted within the first three days of hatching, so contact with the keeper during that time frame is vital if you're to develop a good bond with your birds. Chicks hatched with ducklings will often adopt very duck-like habits, and you'll have difficulty integrating them with other chickens.

The breeds

Pure breeds are those breeds of poultry which, when mated together, produce offspring possessing the same characteristics as the parents which define their type. These breed characteristics are determined and upheld by the Poultry Club of Great Britain, who, in conjunction with the breed clubs, ensure that there is a definitive list of features specific to each different breed. A pure breed chicken is no different to a pure bred dog in that it has been specifically engineered to exhibit certain traits and features, which as a result of selective and dedicated breeding programs, have been set into its genetic makeup.

Ancona

Origin	Italy
Classification	Light, soft feather
Size	Large fowl and bantam
Function	Layer
Egg colour	White through cream
Temperament	Active, shy

Typified by its mottled feathers, giving an overall white spotted appearance on a beetle green background, the Ancona gets its name from Ancona province in Italy. The red comb may be single or rose and in the female will fall to one side of the face and is complemented by neat earlobes that should be white. The legs are mottled yellow, with four evenly spaced toes. A shy and aloof bird, it requires gentle handling and high fencing, although it will produce a good quantity of eggs as a reward for your efforts and can become tame to a degree.

MR P. SMEDLEY

Araucana

Origin	Chile
Classification	Light, soft feather
Size	Large fowl and bantam
Function	Layer
Egg colour	Blue or green
Temperament	Hardy, placid

Famous for its coloured eggs. The eggs of any breed with Araucana in its ancestry will take on a green/blue hue, as this is a dominant gene, and unique in that the colour goes right through the shell. Named after the Indians of Arauca province in Northern Chile, the Araucana is a placid but nonetheless active bird able to be kept both in a closed run and free-range. Colours include lavender, blue, black-red, duckwing, pyle, crele, spangled, cuckoo, black and white. All Araucanas possess muffling to the face and a neat crest, although this is not always passed on to the progeny in a consistent way.

Barnevelder

Origin	Holland
Classification	Heavy, soft feather
Size	Large fowl and bantam
Function	Layer
Egg colour	Dark brown
Temperament	Robust, placid, easily tamed

Originating in the Barneveld district of Holland, these lay dark brown eggs which are one of the main attractions to this robust breed. Being poor flyers, due in part to their heavy nature, the Barnevelder is easily contained behind a medium-height fence, and is at home either in a run or free-range. Allegedly developed from, amongst others, the gold laced Wyandotte, Croad Langshan, Cochin and Brahma, a good quality show bird will have stunning feather colour, making it a very attractive bird for exhibition and garden alike. Colours include black, double laced (black with a red-brown edging), partridge and silver. Prone to Marek's disease, this breed is best vaccinated as chicks.

KAREN POWER

Brahma

Origin	Asia (India)
Classification	Heavy, soft feather
Size	Large fowl and bantam
Function	Ornamental
Egg colour	Tinted
Temperament	Sedate, easily tamed

The King of Chickens, the Brahma was developed originally in America from the Asian Shanghai fowl and Indian Grey Chittagong, and there was much competition and debate between breeders until the official name of Brahmapootras was adopted (later shortened to its current spelling). Nine birds were imported to the UK in 1852 and presented to Queen Victoria. Upright and sedate birds, Brahmas are imposing in a garden setting. Though they possess the heavy leg and foot feathering typical of Asiatic breeds this does not seem to interfere with their otherwise fairly active nature. Colours include dark, light, white, gold, and buff Columbian. Easily tamed, they are ideal in the garden, as a low fence will contain them and they are very tolerant of each other, the cocks being relatively quiet (although by no means silent). Hens will lay a good quantity of small eggs throughout the winter when other breeds are not laying. The only downside to them is that, being large, they aren't particularly suitable around small children, and do need quite a bit of room.

KAREN POWER

Cochin

Origin	Asia (China)
Classification	Heavy, soft feather
Size	Large fowl
Function	Ornamental
Egg colour	Tinted
Temperament	Sedate, easily tamed

An immense bird with a deep shape accentuated by feathering, the Cochin was originally a dual-purpose bird, being excellent for table and a profuse layer. Decades of breeding for show, however, have unfortunately meant that the very characteristics favoured for exhibition, such as their profuse fluffy feathers, have rendered most modern-day lines unable to provide many eggs. Imposing in the garden, colours include black, blue, buff, cuckoo, partridge, grouse and white, and the legs and feet are profusely feathered. Easily kept in little space, fences need not be high as they don't fly well. Their weight also means that perches should be kept low.

Dorking

Origin	Great Britain
Classification	Heavy, soft feather
Size	Large fowl and bantam
Function	Table
Egg colour	Tinted
Temperament	Sedate, reasonably easy to tame

The Dorking (also known as the Darking) is one of the oldest British table breeds, having a lineage dating back to Roman times. Distinguished by its five toes, this bird has been described as quiet and stately and is recognisable by its rectangular shape when viewed from the side. Colours include cuckoo, dark, red, silver grey and white. Needing a fair amount of exercise to prevent them from becoming fat, these birds are easily influenced by improper handling but with care can become relatively tame.

VICTORIA ROBERTS

Dutch Bantam

Origin	Netherlands
Classification	True bantam
Size	Bantam
Function	Layer
Egg colour	Tinted
Temperament	Active, easily tamed

The National Dutch bantam breed, the smallest of all the bantams, but still lay a (relatively) decent-sized egg. Needs plenty of room to be kept busy, but will be quite happy even if kept in a smaller space. The males can be shrill and persistent crowers. Colours include partridge in gold, silver, yellow, blue/silver, blue/yellow, blue, cuckoo (crele) and red-shouldered white (pyle) as well as black, white, blue, cuckoo, and lavender – testament to its popularity. These are a good breed for children as they're easily tamed and often learn to come to hand for feeding.

Hybrids

Origin	Various
Classification	Hybrids are not standardised
Size	Large
Function	Layer
Egg colour	Varies from white to brown
Temperament	Very docile, easily tamed

Probably the most well known of all the laying birds, these are the typical brown layers which inhabit battery farms and are often seen in farmyards, commercially bred for increased egg production utilising as little feed and space as possible. Hybrids bred for free-range are typically more robust and hardier than those bred for cage production. Hybrids are the best option if you're new to chickens as they've been bred with a number of characteristics that are ideal for those just starting out in this hobby:

- Birds are robust and able to stand the rigours of the UK climate.
- Healthy birds are sold fully vaccinated.
- Bred to be docile and easily handled, they have a calm temperament.
- Excellent egg-laying ability, usually in excess of 280 in the first full productive year.
- Rarely go broody.

- Cost effective, as they're relatively cheap to buy.
- You only buy hens as the chicks are sexed at hatching.

Birds are usually upright and sturdy, and come in all sorts of colours although speckled, brown, and white are the most popular.

Bred by crossing monitored and assessed pure breeds with an excellent egg production record, recognised hybrids can only be bred under licence and sold on through approved agencies or stockists, which ensures that the commercial suppliers work hard to promote their good name and stock, much in the same way as breeders of exhibition stock do. By buying from a hatchery you're guaranteed a healthy bird that will produce a large number of fresh eggs during its productive lifespan, as the breeders have invested a lot of time and money in producing their birds and buying the franchise, which is something you can't guarantee if you buy from an unknown supplier.

Unfortunately a number of the large UK hatcheries are rather poor on customer service, so if you prefer more personal attention a better option – if you want their particular line of hybrids – is to buy from a local agent, who you'll find advertised on websites and in the local press. These agents are very helpful and you can often visit their premises to look at the birds. Farmers markets are also a good place to get birds, especially if you can meet the breeders and get some background information on laying ability. Buying from the roadside is *not* recommended.

See the following page for well known and reliable hybrids.

BLACK ROCK

A placid, extremely hardy bird, predominantly black with copper neck markings, and capable of laying around 280 brown, thick-shelled eggs per year. A registered hybrid, only available through agents, so if you see them for sale in a market it's unlikely that they'll be the real thing. At the end of the laying period you'll get a well-fleshed bird for the table.

SPECKLEDY

A speckled hen laying 260 to 270 brown, sometimes speckled eggs per year. Placid and friendly, a good all-round chicken.

WHITE STAR

A white leghorn type, typically laying around 300 to 320 pure white eggs a year. A flighty bird and not long-lived, as all of its energies go into laying.

BLUEBELL

A placid grey bird capable of laying around 260 eggs a year. Bred and imported from the Czech Republic, only agents of Meadowsweet poultry are able to use the name; so, like the Black Rock, if you see them advertised for sale on the roadside then they're unlikely to be the real thing.

CALDER RANGER/ISA BROWN

A brown layer typical of the ones seen on free-range farms. Will produce 300 to 310 eggs in her first full laying year, and is also suitable to be kept indoors, where egg production will be slightly more at 315–20. 'ISA' stands for Institut de Sélection Animale.

Japanese Bantam (Chabo)

Origin	Japan
Classification	True bantam
Size	Bantam
Function	Ornamental
Egg colour	White to cream
Temperament	Placid, easily tamed

Japanese Bantams are truly one of the ancient breeds and have no large fowl counterpart. They have the shortest legs of all the breeds and can be normal-feathered, Silkie-feathered or frizzle-feathered. The large comb of the males, upright tails held at 90° to the body and waddling gait make them a very amusing sight in the garden, where they've been compared to galleons sailing across the lawn! The neat wings are held low, and this combined with the short legs invariably mean that the wing tips touch the ground.

Being a trusting bird they're ideal if you want a breed that children can look after. They're also a beautiful exhibition breed and will lay well, but if you intend breeding them for show then you'll need to bear in mind that the gene for short legs is lethal in double doses, and embryos that receive the short-legged gene from both parents won't hatch but will invariably die in the shell. Overall, 25% of embryos receive the lethal double dose of short-legged genes; 50% receive one short-legged gene and one normal and will hatch as short-legged birds, as this is the dominant gene; and 25% will receive double normal-legged genes and will develop long legs, which are undesirable and will detract from the look of the breed. If two such birds were mated their offspring would likewise have long legs.

There are a wide range of colours available including black-tailed white, buff Columbian, white, black, grey, mottled, blue, cuckoo, red tri-coloured, black-red and various duckwing colours.

Cream Legbar

Origin	Britain
Classification	Light, rare
Size	Large fowl and bantam
Function	Layer
Egg colour	Blue, green, olive
Temperament	Active, reasonably easily tamed

The Cream Legbar was originally bred to enable sexing of chicks at hatching, as the females possess a well-defined pale head spot and are instantly recognisable. In males, the head spot is rougher and spreads all over the body, giving them a lighter appearance. The cream legbar is a crested breed that has Araucana in its ancestry and lays beautiful coloured eggs in hues of blue through green. A sprightly bird, the males will show sparse dark grey barring in the hackles with dark barring on the body, while the females have a delicate salmon breast and softer barring on the body.

Marans

Origin	France
Classification	Heavy, soft feather
Size	Large fowl and bantam
Function	Layer
Egg colour	Dark brown
Temperament	Active, aloof

The original 'speckled hen', this breed from the French village of Marans is famous for its glossy dark brown eggs, which are very desirable. Unfortunately exhibition lines tend to lose the egg colour in favour of show markings, while a line of good brown egg layers will show too many faults in type to be successfully exhibited. Active but wary birds, they respond well to calm handling but never become fully hand-tame. Colours are black, cuckoo, golden cuckoo, and silver cuckoo.

Marsh Daisy

Origin	Britain
Classification	Light, rare
Size	Large fowl
Function	Layer
Egg colour	Tinted
Temperament	Active, aloof

Created in Southport in the 1880s by crossing an Old English Game bantam cock with Malay hens and then mating a cock from this mating with black Hamburg/white Leghorn cross hens. A resulting cock was then crossed back to the Hamburg/Leghorn hens and the offspring line bred until around 1913, when Pit game and Sicilian Buttercups were used to give us the bird we recognise today. Classed as a rare breed, the Marsh Daisy is seeing a resurgence of popularity amongst people who wish to preserve the old English breeds. Marsh Daisy colours are black, brown, buff, wheaten and white.

KAREN POWER

Modern Game

Origin	Britain
Classification	Hard feather
Size	Large fowl and bantam
Function	Game
Egg Colour	Tinted
Temperament	Active, easily tamed

Easily recognised by its upright stance and long legs and neck, this breed was developed in England to replace fighting birds after the ban on cockfighting. Bred now for exhibition, Modern Game are poor layers but do make a placid pet if you find their shape attractive. The practice of 'dubbing' whereby the comb is removed after hatching is now actively discouraged. Thirteen colours have been standardised, including birchen, black-red, brown-red, golden duckwing, silver duckwing, pyle, wheaten, black, blue, white, blue-white, silver-blue, and lemon-blue.

JENNI O'SULLIVAN

Orpington

Origin	Britain
Classification	Heavy, soft feather
Size	Large fowl and bantam
Function	Layer
Egg colour	Tinted
Temperament	Sedate, easily tamed

Originating in Orpington, Kent, in the late 19th century, this was a dual-purpose breed with Langshan in its ancestry, amongst others. Layers were said to be able to produce over 300 eggs a year, which would have made it *the* egg producer of its time. Unfortunately, like so many other breeds shown today, the Orpington is now unlikely to lay more than about 50 eggs a year, due in part to the development of under-fluff and loose, soft feathers for exhibition. A docile bird, they're a familiar sight at many shows and recognisable by many people. Their large stature makes them a good garden bird, as they don't fly and can be restrained by low fences around borders. Their only requirement is a dry area to perch, to prevent their under-fluff from getting waterlogged. Standard Orpington colours are blue, black, white and buff.

Old English Game

Origin	Britain
Classification	Hard feather
Size	Large fowl and bantam
Function	Game
Egg colour	Tinted
Temperament	Active, easily tamed

Another breed originally created specifically for fighting. There are two types of English Game, the Oxford and the Carlisle, the latter having a broader breast. There is still an active trade in these birds, and unfortunately if you decide to keep them heightened security will be needed to prevent theft. The Carlisle comes in 15 colours, the Oxford in around 30. Not a particularly good layer, but a friendly bird and a good choice if you want a breed that children can look after. Although the cocks may fight amongst themselves the hens are tolerant of other birds.

Pekin

Origin	China
Classification	True bantam
Size	Bantam
Function	Ornamental
Egg colour	Tinted
Temperament	Active, easily tamed

Although it's also known in Europe as the Cochin Bantam, the Pekin is not really a bantam version of the Cochin, as it is too dissimilar. A round and cuddly ball of soft feathers, this must be the children's favourite as it's easily tamed and will run towards you in excitement if it thinks a treat is in store. The hens lay few eggs and go broody very easily, making them useful for hatching the eggs of non-sitting breeds. You'll need to provide a dry perching area, as their foot feathers are easily soaked and soiled in the rain. Although normally placid, Pekins can act violently towards other birds and will need separating if this occurs. Colours include black, blue, buff, cuckoo, mottled, barred, Columbian, lavender, partridge, and white. Other colours do exist but have not been standardised, for example the lemon cuckoo.

Poland

Origin	Poland
Classification	Light, soft feather
Size	Large fowl and bantam
Function	Layer
Egg colour	White
Temperament	Active, relatively easily tamed

An old breed, although its ancestry is unclear due to many countries having laid claim to its making. Recognised instantly by its massive crest, the exhibition Poland is truly a sight to behold. This is a fairly high-maintenance breed, with a number of conditions being made worse by its crest, such as lice infestations and eye problems, although there is movement within the breed club both at home and abroad to reduce the crest to a more practical size and shape. Bantam Polands do make good pets and will, like Pekins, run to their handler for food. Ensure that drinkers and feeders don't allow the birds to soil their crests, which can be taped up or cropped if the birds are to be kept outside to free-range. Standard colours include chamois (buff laced), gold, silver, self-white, self-black, self-blue, white-crested black, white-crested cuckoo, and white-crested blue. In birds with larger crests the head is very delicate, and if alarmed – particularly if approached from above – the bird will fly upwards and knock itself out.

Rhode Island Red

Origin	United States
Classification	Heavy, soft feather
Size	Large fowl and bantam
Function	Layer
Egg colour	Mid-brown
Temperament	Active, easily tamed

Originating in Rhode Island State in the USA, this breed must lay claim to being one of the most popular of all time. Used extensively in hybridisation, due to the ability to identify its chicks by sex linking. This and the fact that its brown eggs are much favoured in the kitchen mean that the Rhode Island Red was originally bred as a dual-purpose chicken. Shaped like a long rectangle and often referred to as 'brick-shaped', it is typified by its beautiful mahogany plumage and a calm friendly nature that makes it ideal for the beginner. Although suitable for an enclosed run, its eggs improve in colour and flavour if the birds are allowed to free range. This will also prevent fighting amongst the males.

Sebright

Origin	Britain
Classification	True bantam
Size	Bantam
Function	Ornamental
Egg colour	White through to cream
Temperament	Active, easily tamed

Made in the early 1800s by Sir John Sebright to create a laced bantam, the Sebright is unusual in that the male has similar feathering to the female and is said to be 'henny-feathered', lacking the typical male plumage; which as a consequence means both males and females are beautifully laced all over. Active and inquisitive, Sebrights make good tame pets and can be kept in a relatively small space, although you'll not get many eggs. Males are tolerant of each other and not particularly shrill or noisy. Unfortunately the Sebright is susceptible to Marek's Disease, so you'll need to ensure stock is vaccinated unless you prefer to breed for resistance. Colours are silver laced and gold laced.

KAREN POWER

33

Silkie

Origin	Asia
Classification	Light, soft feather
Size	Large fowl and bantam
Function	Ornamental, sitter
Egg colour	Tinted through to cream
Temperament	Docile, easily tamed

The Silkie is instantly recognisable by its pompom, fifth toe, black skin and profuse fluffy plumage, caused by an absence of barbs on the feathers that prevents them from knitting together. Known for its ability to brood the eggs of other birds, although the chicks can get caught up in the under-fluff and die, so a Silkie cross with harder feathers is better. It is docile and the cockerels placid and not particularly noisy. Easy to hand-tame, Silkies don't fly and are easily confined behind a low fence. Hens come back into lay around Christmas-time so are reliable when other breeds are still off lay. Unfortunately susceptible to Marek's Disease, so you'll need to ensure stock is vaccinated. Colours include white, black, blue, partridge, and gold.

Sultan

Origin	Turkey
Classification	Light, rare
Size	Large fowl and bantam
Function	Ornamental
Egg colour	White
Temperament	Placid, relatively easily tamed

A truly ornamental breed descended from a group of birds imported by Elizabeth Watts in 1854, the Sultan is not particularly popular but does have a dedicated following. The breed standard lists a crest, beard, vulture hocks, feathered legs, five toes and white plumage as attributes – features that in many other breeds would be classed as faults. Placid and good-natured birds, they should be provided with an enclosed and sheltered run, with specialist drinkers to prevent the beard becoming soiled. The only recognised colour is white.

Sussex

Origin	Britain
Classification	Heavy, soft feather
Size	Large fowl and bantam
Function	Table
Egg colour	Tinted
Temperament	Docile, easily tamed

Bred originally as a table bird, hens will also lay a quantity of good-quality eggs, which makes it a useful dual-purpose breed for the kitchen garden and also commercially, as Sussex crosses make fine producers of both meat and eggs. Calm and docile, it's a good chicken for the beginner and easily tamed. Colours are speckled, brown, buff, light (Columbian), red, silver, and white.

Transylvanian Naked Neck

Origin	Romania
Classification	Heavy, rare
Size	Large fowl and bantam
Function	Ornamental
Egg colour	Tinted
Temperament	Friendly, easily tamed

Instantly recognised by its lack of neck feathers, the Naked Neck is notable for its inherent reduced level of abdominal fat, making it an excellent choice when crossing for meat birds. Chicks are hatched without neck down, making them appear very delicate and weak, but don't be fooled, as this is actually a remarkably hardy bird bred to withstand a cold European climate. The hens are good layers and will get most of their feed from free-range if allowed. Colours are black, white, cuckoo, buff, red, and blue.

Welsummer

Origin	Holland
Classification	Light, soft feather
Size	Large fowl and bantam
Function	Layer
Egg colour	Brown to deep brown
Temperament	Active, aloof

Famous for its dark brown sometimes speckled eggs, the Welsummer has been bred as a layer, and the hen's prowess can be seen in the fact that her legs will pale from yellow to white as the season progresses and the colour is drained from the leg pigment to colour the yolk. If breeding you should select eggs of a lighter brown colour, as this denotes good laying ability due to the fact that their colour pales after several weeks of laying; darker eggs are consequently from hens which lay considerably less. Although partridge is the most common accepted colour, there is also a silver duckwing.

Wyandotte

Origin	United States
Classification	Heavy, soft feather
Size	Large fowl and bantam
Function	Layer
Egg colour	Tinted
Temperament	Friendly, easily tamed

A heavy utility breed from America that was allegedly named after a ship. The body should be broad and round with profuse feathering, and from the back the tail is an upturned V. The backside and breast of the hens should be deep and rounded, indicating laying prowess. Once her clutch is laid the hen is reliably broody, although you'll need to take care of her during this period as she'll often not get off the nest to feed. Not great flyers, Wyandottes make excellent free-range pets and are an ideal breed if you're considering exhibiting. Colours include barred, black, blue, buff, Columbian, partridge, silver pencilled, red, silver laced, gold laced, blue laced, and buff laced.

Getting your chickens

So, having decided that you want to keep chickens, the next question is where do you get them from? The livestock sections of newspapers invariably include ads placed by people seeking to rehome unwanted birds. However, unless you know the seller either by direct contact or reputation this route isn't recommended, as you don't know what problems you may be buying in along with the birds.

Hatcheries and their agents

A better option is one of the national suppliers, who, being large commercial organisations, can supply their customers with a large variety of healthy pullets (female birds under one year old) in small or large quantities. They're ideally suited for the free-range market and can often supply organic 'point of lay' birds (see page 184) if requested, which are reared to Freedom Food Standards, are fully vaccinated, and have full traceability.

For best service use one of their local agents, as these are better situated to provide a friendly and efficient service to customers. The birds that the agents sell are hybrids, which can satisfy a variety of your preferences and can lay in excess of 300 eggs per year during their first full laying season. Many agents also offer other types of poultry, such as traditional table chickens and turkeys (including Bronze turkeys) that dress out at 4.5–25kg (10–55lb).

Below: A breeding group of Norfolk Black turkeys.

JANICE HOUGHTON-WALLACE

Specialist farms

With years of expert knowledge about poultry and wildfowl, many animal farms and parks not only offer expert advice on care but can also help you choose the right breed for your needs, as well as supplying all the necessary feed, housing, and equipment. Often nationally recognised Centres of Excellence for a wide variety of wildfowl, many are actively involved in different areas of wildlife conservation.

One particular advantage over the national suppliers is that you can go along and have a look at the different breeds to get a better idea of what they actually look like.

Poultry markets and auctions

Markets are a more traditional source of birds, and many offer two distinct types of poultry sale: the regular 'fur and feather' sales, and the more interesting rare breeds sales which are often held twice a year, once in the spring/early summer and again in the autumn.

With the rare breeds sale, all entries are examined and graded by breed specialists prior to the auction in order to maintain the highest quality of stock on offer, thus giving you a reliable idea of the birds' health and pedigree. If you're looking for a laying hen then the fur and feather sale is a much more suitable option.

Rescue

The British Hen Welfare Trust was the UK's first registered charity for laying hens, and was established by its founder, Jane Howorth, in order to raise awareness of the 20 million hens being kept in cages in this country. She began rescuing battery hens from slaughter for her own peace of mind, and several hundred hens later decided to take 100 hens from a local battery farm in Devon for the specific purpose of finding them good homes. Since then tens of thousands of hens have been rescued from slaughter and given the opportunity to enjoy a free-range retirement, with many going on to become much-loved family pets.

Unfortunately many less ethical farmers have caught on to this growing source of revenue and will offload caged laying hens past their best on to unsuspecting owners, who are consequently greatly upset when the birds die or never lay an egg. So if you're going to rehome

Above: Battery hen rescue can be rewarding.

birds, get them from a recognised charity to ensure you get the best information and support.

Do give consideration to the fact that rescued caged commercial hybrids, with their weak feathering, poor immune systems, and high food demands, do not thrive in exposed outdoor situations and aren't recommended if you want a lot of eggs. These ex-battery chickens, which have known only a completely protected environment for their whole lives, are quickly stressed if allowed extensive free-range and are much better in the back garden, where exposure to the elements is limited and they have the security of a nearby shelter. Unfortunately, because they're stoic and tolerate whatever conditions we subject them to, their unhappiness won't necessarily be obvious, and they'll go on producing eggs until a stress-related illness kicks in and the bird suffers.

Hatching eggs

The most educational and interesting way to obtain chickens, if you're an experienced keeper and have the necessary equipment, is to buy eggs from an external breeder and hatch your own, as there's nothing more magical for children than watching that first chick break free.

This option is also a favourite with keepers who're not in a position to use their own birds, perhaps because the close vicinity of neighbours or the existence of deed clauses mean that they can't own a male. Others may want a new bloodline without the hassle of travelling for adult birds, or even need to

Above: Hatching your own eggs is educational and interesting.

restock after a particularly virulent outbreak of disease, when eggs offer sufficient protection from the transmission of infection.

So consider this: eggs are the ideally packaged form, you don't have to feed or water them, and you can get hundreds in your car if need be. Fortunately, in today's technological society you don't even have to leave your house, as the Internet offers a practical and efficient solution through auction sites such as eBay.

By following a few simple guidelines, you'll be able to buy your eggs with confidence:

1 Research your intended breed and know its standard – that way you'll know if the eggs being offered have the potential to grow into the correct type, and be cautious if no picture is offered. Always check with the seller that the image they've shown you is of their own birds, and whether or not the eggs are from the birds pictured.

2 Read the advertisement copy thoroughly, and if there's anything that you don't understand, or need clarifying, *ask* – a good seller will always reply and won't berate you for a simple question, as everyone has to start somewhere. Ask plenty of questions. If the ad says 'show quality', ask the seller which shows they attend and their best results that season; ask too if they're a member of the breed club, because this will demonstrate that they should know their standards and be a responsible breeder, as many of the clubs demand that their members don't bring the club into disrepute. Be warned, though, that just because the parents may be good quality, it doesn't necessarily follow that the offspring will be – it's always worth checking on the seller's feedback and contacting past buyers to see how they got on.

3 Depending on what you want your stock for, take statements such as 'unrelated birds' with a pinch of salt, as this isn't necessarily a good thing if you want the specific line (based on the information that you've gleaned about the seller's birds). Many poultry exhibitors and breeders 'line breed' from an established genetic pool of related birds, as unrelated birds can add detrimental genes to exhibition stock. Conversely, though, unrelated stock can add beneficial vigour to a line suffering from too much inbreeding, and can even improve the stock. So this is a decision that has to be based on your own preference.

Below: Breeders use established genetic pools to breed identical birds.

4 Look out for misleading advert statements such as 'L@@K RARE!!!!!' Again, if you know your standards you mustn't let yourself be persuaded into buying something that, due to the nature of its genetics, will die just before or shortly after hatching. A true rare breed is listed as such in the British Poultry Standards, while a rare colour is often one that's not standardised and can't be shown at many exhibitions.

5 Pay attention to the pictures in the advertisement. Do the birds look in good health? Is the surrounding area clean or muddy?

6 Once you're happy that you want the eggs, check the posting details. eBay insists that hatching eggs are posted via a next day service, so all sellers should offer that. Don't be tempted to go for the cheaper First Class mail option, always use Special Delivery as that way the eggs are kept out of the automatic sorting tumblers and there's less chance of damage to the embryos or their air sacs. The price for this service is around £7 to £10 at the time of writing, depending on the weight of the parcel, but it's well worth paying. Check the packing details too: polystyrene boxes offer good protection but aren't absolutely necessary if the seller packs securely in normal egg boxes and secures these within an outer carton.

The main advantage with polystyrene is that the eggs are cushioned, protected from major temperature variations, and secured within the parcel, which prevents movement and so avoids displacement of the air sacs and internal fluids, which can reduce the eggs' viability. Despite some sellers' claims to the contrary, polystyrene boxes don't adversely affect fertility, which is more likely to be the fault of the seller.

Below: Packing should provide ample cushioning.

7 Always stick to your limit when bidding, and don't get caught up with the excitement and bid more than you intended. Often a seller will send a 'second chance offer' to the next highest bidder, and most relist within seven days anyway, so if you miss one lot there's always another chance a little later.

8 Having won the eggs pay immediately, as time is critical if you want to get the freshest eggs delivered with the minimum possible delay. Ask when the seller is posting them and ensure that they inform you once the eggs have been dispatched so that you can be at home when they arrive.

9 Once you've received your eggs, check them for quality and damage and contact the buyer *immediately* if you're unhappy. Always candle your eggs on receipt (see page 112), to check for partial incubation (recognised by a blood circle) or hairline cracks. Check, too, that the postal service you paid for is the one by which they've been sent.

Below: A candling lamp will highlight any hairline cracks.

10 Some people will then let the eggs settle for 12 to 24 hours before setting (see storage page 111). Know your incubation techniques, and don't expect a 100% hatch, as eggs are a natural product and subject to all sorts of problems with development – so read the seller's feedback to assess how good their hatches are, and don't rely on the percentage score alone as follow-up, which often contains more information, isn't counted.

Chicks

Day-old chicks have the capacity to go several days without food, as they still have the remnants of egg yolk sealed inside their bodies. This makes transport easier, as it means they can travel for several hours as long as they have warmth and access to water. Traditionally day-olds were shipped across the country via the rail network, with boxes being collected from depots and stations by the new owners. With the demise of the railway networks, however, modern options are Royal Mail Special Delivery or courier, both motorcycle and car.

Above: Chicks can survive for about 48 hours on the absorbed yolk.

Table birds are normally purchased as day-olds and then grown on for the table, some breeds being ready for slaughter by 16 weeks. Hybrid chicks will normally have been sexed at the hatchery, although you shouldn't expect 100% accuracy as this is a difficult process and you may sometimes find a male amongst the females. You'll need to provide a sheltered area free from draughts and protection from rats, along with a heat source. Traditionally a brooder is used, which is an enclosed area that doesn't allow the chicks to wander too far from warmth.

Chicks will need a 'chick crumb' ration. This may or may not contain a coccidiostat, which has anti-coccidial properties to protect growers from developing coccidiosis (see section 5) by providing low-level exposure that allows them to develop immunity.

When the chicks arrive, remove them carefully from their carry box to the brooder – don't be tempted to just tip them into it as shown in certain television productions, as you'll increase their stress levels and may cause injury; such actions are thoughtless and cannot be condoned. The brooder equipment should be set in a vermin-proof house and can comprise either a heat-lamp arrangement or an enclosed brooder unit. It should be set at a temperature of 25–35°C, depending on the age of the chicks, so ensure your thermometer works before the chicks arrive:

Age	Temperature
1 day	35°C
7 days	33°C
14 days	30°C
21 days	28°C
28 days	25°C

Make sure that a chick feeder filled with chick crumb is available, along with a chick drinker filled with fresh water to which a probiotic has been added.

Above: Growers should be secured against predators.

Growers

These are youngsters, usually from the age of six weeks, and with them the most demanding process of hatching and brooding has been done for you. They'll be fully feathered and hardy, that is ready to go outside. Depending on the weather, though, you may still need to provide a heat source, especially with less hardy breeds, while older birds of 7 to 12 weeks will benefit from free-range in a run, with a house to protect them from the elements and predators. The survival rate of growers is higher than that of day-olds, although the initial cost per bird will be more. Specialist growers' pellets are available, which are higher in protein content than ordinary layers' pellets.

Growers, like other new birds, should be introduced quickly and quietly to their new accommodation, which needs to be rat-proof to prevent injury to the new stock. Ideally the house should be enclosed and moveable so that the young birds have access to grass during the day and shelter when it's needed, before returning to the safety of the house at night. A perch should be provided, which the youngsters will quickly learn to use.

Point of lay

Females can start to lay eggs about 16–20 weeks after hatching, and are then described as being at 'point of lay'. Reliability is the main factor when buying point of lay (or POL) birds, since they've nearly reached adulthood and have survived the worst of the ailments that can affect youngsters. The cost of such birds is higher, but you can start to get eggs almost immediately, you haven't had to pay feed costs while the birds were growing, and, naturally, all the birds will be female.

Should you decide that adult birds are needed, point of lay pullets can be bought in that will come into lay while your older hens are having a rest. When your birds arrive, they'll be stressed and hot, and may be sneezing or look depressed. Don't be alarmed, instead quickly and quietly move them to their new quarters that will have been prepared in advance. If you have children, it's best to keep them away from the birds during arrival as their eagerness can alarm the pullets further. Place the birds into their house through the largest opening (usually the back door or hinged roof) and leave them together with a filled feeder and drinker until morning.

Below: Point of lay pullets emerge the following morning.

Above: Delivery of pullets will be arranged by the supplier.

Above: Birds should be placed directly into their quarters.

Above: The new birds will be nervous of their new surroundings.

Although there are a number of valid reasons why it's inadvisable to mix birds of different ages (see box), this may not be practical in today's society, with its smaller gardens and restrictions on the amount of available space. Consequently you may encounter problems when introducing new birds into an established group. New chickens may well be picked on for a few days, but harmony returns once the older layers have re-established their

Age differences

Reasons for keeping chickens of different ages separate

- Some diseases are age-related and either transmissible or infectious at certain stages in development. They may also be lethal to one age group and not to another.
- Older birds are more likely to be carriers of disease.
- Older birds will bully younger birds.
- Younger birds (*ie* less than a year old) should not be mated.
- Feed requirements are different.
- Care requirements may differ – for example, some groups will require more or less heat provision or light provision.

pecking order. You just need to make sure you're on hand should the fighting become more serious and wounds arise, since the sight of blood will mean all the birds suddenly become more interested in events, and fatal injuries can result from aggressive attacks.

Sometimes the birds will keep on at each other until they become absolutely exhausted, making them an easy target for a determined troublemaker. If this happens (or if you want to do so from the start), separate the offenders but if possible keep them in sight of one another, to reinforce the hierarchy – you can put a small house and run inside the main run, for instance, with the new chicken inside it, so that the others get used to her.

It is recommended that you introduce new birds at night when all is quiet. Then when the others wake up with the newcomers already in their midst they may well be accepted much more readily. *Never* introduce a solitary bird, as she's sure to be attacked – remember that there's safety in numbers. Keep gentian violet spray handy too; it disinfects wounds and colours them purple, making them less attractive to would-be peckers.

Below: Gentian Violet spray will mask wounds and is mildly antiseptic.

Other poultry

DEFRA defines poultry as comprising any bird in the following groups:

- Chickens (including bantams)
- Ducks
- Geese
- Turkeys
- Quail
- Guineafowl
- Pheasants
- Partridges
- Pigeons (reared for meat)
- Ostriches
- Emus
- Rheas

You may find that, following your introduction to chickens, you also become interested in one of these other types of poultry as your involvement in the hobby progresses; or it may be that one of these other varieties is your main interest anyway. Either way, just as with chickens you'll need to research your particular breed by contacting as many breeders and reading through as much literature as you can.

Ducks

Originally kept as a source of eggs and meat, the modern-day duck has also been bred to fulfil another market requirement, that of visual pleasure. Commercial poultry production was set up on a large scale following World War Two, and as a result chickens, ducks, and geese all but vanished from gardens and small farms, as it was far cheaper and easier to buy from the larger facilities. More recently, however, the efforts of specialist breeders under the guidance of the British Waterfowl Association (BWA) has resulted in an upsurge of interest and the preservation of many rare and endangered breeds as people try to recreate an idyllic image of farm life.

Ducks are a familiar sight on many rivers and ponds, and can be a particularly attractive addition to a garden setting, although it should be noted from a practical point of view that they can cause damage in smaller gardens due to their habit of mixing their food in mud and water before eating it, creating puddles (hence the term 'puddling') and giving rise to the dabbling noise we associate with them. However,

Below: Ducks represent an idyllic lifestyle.

Above: Ducklings automatically command your attention.

Above: An old bath serves as a pond and helps recycling.

they're invaluable as pest destroyers, and have the distinct advantage over chickens of not indiscriminately scratching up everything in the area in which they're foraging, preferring instead to use their specially adapted bills to feel what they're eating – if you can train them to eat corn from your hand you'll appreciate the flexible and rubber-like nature of this specialised tool.

Requirements

Like all other poultry, ducks must be protected from predators – especially if you have a pond, as they'll sleep in the middle of it and make themselves an easy target, literally a 'sitting duck'. This can easily be solved by training the ducks to go into their house at night – walk slowly behind them with your arms outstretched and use a vocal command, effectively herding them along.

Housing can be as simple as you wish, with about a square metre (roughly 10.75sq ft) of floor space needed for a trio of small ducks such as call ducks, or 1.3m² to 1.5m² (approximately 14–16sq ft) for larger breeds, lined with a thick layer of fresh straw for bedding. Cleaning is a simple affair as you'll just need to remove the top layer of compressed straw which has been patted down by their webbed feet, by peeling it off and replacing with fresh, probably twice a week during wet weather. As far as the enclosure is concerned, most of the domestic breeds don't fly so a fence about a metre (3.3ft) high will keep them in.

Being waterfowl, it's essential to provide them with permanent access to water to enable them to swim and feed. The smaller breeds will manage with a large paddling pool or tub, although like the larger breeds they'll thrive if you can provide a pond.

Ducks are omnivores and, like chickens, will eat both animal and vegetable matter. Wheat and maize can be fed along with bread, and you can purchase a complete duck feed from animal feed stores (although it's not yet as readily available as chicken feed), along with flint grit. Acorns can be fed as a treat.

Small ducks

CALL DUCK

Originally imported from America, this duck should have a rounded head with a short, broad bill. Due to its persistent and distinctive quack, which is easily recognised at any waterfowl exhibition, British keepers bred the call duck to 'call in' wild ducks for shooting.

BLACK EAST INDIAN

A beautiful black-feathered bird that shines an iridescent green when the sun strikes it. Older birds develop white feathers in the same way we grow grey hair. An excellent forager.

Below: Small ducks are as pretty as their large counterparts.

Light large ducks

RUNNER DUCK
Instantly recognisable by its upright stance, the runner duck needs less water and is perfectly happy with a large tub in which to immerse its head and dabble.

KHAKI CAMPBELL
A beautiful brown duck bred by Mrs Adele Campbell in an effort to produce a prolific egg layer. When buying ensure that you have a utility line rather than an exhibition line, so that egg laying isn't reduced, as showing lines of birds have lost much of their capacity to produce numerous eggs. There is also a white variety.

Heavy large ducks
These need deep water to facilitate successful mating. They include:

AYLESBURY
One of the best table breeds, it has suffered from hybridisation in recent years, with many white birds being sold as Aylesburys that aren't. So be sure of your breeder when purchasing.

ROUEN
One of the most striking breeds of duck, originating from France, this huge table bird is subtle in colour and very docile. Like the Aylesbury, it's characterised by its block-shaped body. The eggs are green tinted.

CAYUGA
Taking its name from Lake Cayuga in New York State, this beautiful bird has black feathers with a lustrous beetle-green sheen. When picking breeding birds, it's best to avoid those with a purple or brown sheen as this is an exhibition fault. Like the Black East Indian, the older birds develop white feathers. First eggs may be very dark, almost black in colour, but this fades with subsequent lays.

Below: Cayuga.

Above: Muscovy.

SILVER APPLEYARD
A dual-purpose breed named after Reginald Appleyard, a domestic fowl judge and renowned breeder. This is one of the first breeds of the year to start laying, and has a distinct and palatable flesh as a table bird.

PEKIN
Sleeker and lighter in weight than the Aylesbury and originally bred in China, this upright, chubby bird is very distinctive and suitable as a table bird. Fertility problems may be encountered if deep water isn't provided for mating.

MUSCOVY
Although more goose-like, the Muscovy is a consistent grazer and used as a sitter to hatch ducklings, the duck equivalent of a broody hen. If a Muscovy drake is mated to any other duck breed, the offspring are sterile. Muscovy ducks have very sharp claws, although on the plus side neither sex quacks.

Eggs

Ducks can lay throughout the year, making them a valuable source of eggs over the winter when your chickens are having a rest. Some breeds such as the Khaki Campbell lay well over 300 eggs per year, having been bred for that purpose. Ducks lay normally early in the morning, so you'll need to collect the eggs as soon as you can since they quickly become soiled by the adult birds' watery droppings.

In the past duck eggs have suffered very much from bad press, even so far as being labelled poisonous in order to reduce consumption. The main reason for this reputation is that the shell is more porous than that of a chicken egg and is therefore easily affected by micro-organisms such as salmonella when kept in poor conditions (ducks are inclined to be messy and their eggs are easily soiled). Today's more rigorous hygiene regulations, including vaccination, have more or less eradicated the problem, making duck eggs a flavoursome alternative to those of chickens.

Geese

These are probably one of our most recognisable farmyard animals as a result of the stories we learned during childhood, the goose being a prominent feature of nursery rhymes and fables featuring memorable characters such as Mother Goose and the Goose that laid the Golden Egg. We also regularly refer to geese in conversation, in expressions such as 'cooking your own goose' and 'goose bumps', and as a result our everyday lives contain constant references to this graceful animal.

Geese can live for over 25 years and are rarely troubled by illness, being resilient to many of the problems commonly encountered by other species of domestic poultry. They are kept today as grazers and guards, due to their fiercely protective nature. Geese can be stocked at a rate of up to five birds on a quarter of an acre, as 80% of their diet will come from grass, making them very economical to keep. If you start a group on grass in the winter they'll invariably be on top of the grazing by spring and will save you the job of mowing enclosures for the rest of the year. Being grazers, the goose's specially adapted bill will shear the grass close to the ground, keeping the sward short and encouraging thicker growth.

Geese exhibit a unihemispheric sleep pattern where one side of the brain sleeps while the other remains alert to predators, and they're easily alerted to danger, so that their loud trumpeting will give plenty of early warning should any local fox or other nocturnal visitor – whether on four feet or two – think of interfering with your poultry. Although extremely intelligent fowl and very able to defend themselves, like all other poultry they must be locked up at night.

Requirements

Besides the obligatory security fencing, as a fox is capable of carrying off an adult bird, it's essential to have a solid house that provides shade and shelter from the elements, and into which the birds can be shut up after dark. It should have a floor area of 2m² (21.5sq ft) for a trio of birds, and be about

Below: Geese should have access to water.

Below: Geese are constantly alert to danger.

Above: Geese should be herded in at night.

1m (3.3ft) high. Make sure there's plenty of ventilation (without creating a space through which the birds can stick their heads), a good thick layer of straw for bedding, and an adequate water source, as geese, like ducks, are waterfowl. Geese will learn to be herded, and should be shut in at night to avoid fox predation and theft.

Light breeds

CHINESE
One of the best guard breeds, Chinese Geese are easily recognisable by the knob on their heads. Laying between 60 and 80 eggs a year, they're also the best layers of all the geese.

Below: Embden females.

> ## Fact...
> The skins of goslings were once used as powder puffs.

Medium breeds

ROMAN
A good grazer, this is one of the smallest geese. Pure white with a pink/orange bill, they can chatter a lot, so aren't an ideal choice if you have easily irritated neighbours. A good sitter, this breed will not require a pond for successful mating.

SEBASTAPOL
Characterised by long curly feathers, this is the breed most often chosen for exhibition.

Heavy breeds

TOULOUSE
This grey/white-feathered bird came to England from France in the 14th century. It has a large dewlap (a fold of fleshy skin) under the chin. Gentle and quiet, a true Toulouse goose has a blocky body shape with a deep keel and full paunch. Water is essential for successful mating.

EMBDEN
One of the hardiest of the white breeds, this goose will mature rapidly. A good bird should have no sign of keel, which if present suggests the bird is likely to have Toulouse in its ancestry. As with Toulouse geese, water is required to guarantee successful mating.

Below: Goose eggs are around twice the size of a large hen's egg.

Eggs

Often starting to appear in February and continuing until June, eggs should be collected as soon as possible after laying and will benefit from cleaning in warm water with a sponge scourer. Stronger than chicken eggs, goose eggs are excellent fried, boiled for around six minutes, or made into an omelette.

Turkeys

Descended from the wild North American turkey, the species ranged far and wide over Central and North America and was revered by the Aztecs. Today we're more familiar with the turkey as a table bird for Christmas celebrations, although more recently – thanks to the hard work of the UK Turkey Club – there has been an increase of interest in exhibition birds and year round usage of turkey meat and eggs.

Male turkeys (stags) can be kept with females (hens) at a rate of one stag to 12 hens, which will come into lay at about 36 weeks of age and lay 50–100 creamy coloured eggs per year. These should be collected three times daily to prevent egg eating, and the hens should be protected with a poultry saddle from the amorous attentions of the male, which can cause serious injury to their delicate skin. Worming should be carried out every six weeks to destroy the Heterakis worm, which is the intermediate host for the lethal Blackhead protozoan.

Requirements

A normal small garden shed, 2.4m x 2m (8ft x 6ft), with plenty of straw on the floor for bedding, makes ideal overnight accommodation for up to 8 birds as long as the birds are allowed to free-range during the day. It should be fitted with a perch, 7.25cm (3in) round and about 75cm (36in) from the ground. Birds will need to be driven in at night to stop them trying to roost.

Younger birds (poults) in particular will need secure housing and protection from thunderstorms, as they're less robust than the young of other species and can suffer heart attacks due to high blood pressure if kept intensively.

Like other poultry, turkeys will eat both plants and insects when ranging, and should be provided with a handful of whole wheat in the afternoon. Flint grit or mixed grit should be provided, but never oyster shell on its own, as it may become lodged in the crop. Specialist feed merchants can supply complete turkey feeds for chicks, growers and adults, which must not contain coccidiostats, as these are harmful to turkeys; a chicken ration is unsuitable.

Quail

Quail are related to the pheasant and partridge and in the wild are denizens of undergrowth and grassy fields, which provide them with cover. They're nervous birds and when disturbed will either stand motionless or, if really alarmed, will break cover and fly straight upwards.

Ideally suited to an aviary, they'll range over the floor picking up scattered grains when kept with other birds. Coturnix is the breed kept mainly in the UK, although Chinese Painted and Bobwhite are also popular, and are kept for meat and eggs.

Feeding

They'll eat proprietary chicken pellets quite happily, with added treats such as lettuce leaves, canary seed, and millet. Leaves from the kitchen garden may help to prevent feather

Below: The pied (Crollwitzer) is an ornamental turkey.

JANICE HOUGHTON-WALLACE

Below: Quail.

pecking in groups of birds, as it also provides them with something to do. Although its benefits are debated amongst keepers, you should provide access to grit if feeding grains, as well as a permanent supply of clean fresh water. Cuttlefish may be used to keep the pointed beak sharp, as it's designed to cope with insects.

Feed should be placed in suspended feeders as quail are great scratchers and will throw feed everywhere if they can get into the feed hopper.

Guineafowl

Although related to the pheasant family, guineafowl originated in Africa and are kept as a table bird, having a good abundance of breast meat. Guineafowl prefer to perch in trees and can be a nuisance in the evening, being very noisy as they prepare to roost. The shrieks of guineafowl can annoy neighbours and they're really best kept only if you have a lot of space. They do make ideal guards.

Requirements
You'll need to provide perches.

Feeding
Chicken pellets and grain are ideal.

Below: Guinea Fowl.

Above: Pheasants are mainly reared for shooting.

Pheasants

To keep pheasants, you'll need an aviary of some description since, unlike chickens, game birds will fly away if they're not securely housed. Under the current Wildlife and Countryside laws you must be licensed to release non-native birds into the wild, and you could be fined if you have any escapees.

There are many species of pheasant, some of which are suitable for a beginner and don't require special treatment, heating or food to stay healthy. The Golden pheasant is probably the most popular, along with the various Silver pheasants and the Lady Amherst's pheasant, which is renowned for the male's long tail feathers. There are also various species of Jungle Fowl, although be warned that the males do crow if you're thinking of keeping them in an urban area.

It's always best to buy unrelated stock, so ask the breeder from whom you intend to purchase if they guarantee that the birds are unrelated.

Requirements

In most cases, pheasants are kept in pairs, or sometimes in threes (two females and one male), as normally if there's more than one male in an aviary they'll fight for possession of the female, especially during the breeding season.

In the case of the smaller species, such as the Golden pheasant, an aviary a minimum of 3m (9.8ft) long by 2m (6.5ft) wide, plus a shelter 1m (3.2ft) deep by 2m (6.5ft) wide, will satisfy one pair; for the larger species, the minimum requirement is twice that. The aviary should be at least 2m (6.5ft) high to allow you to walk around inside, and the sides should be covered in galvanised wire of 25mm (1in) mesh to keep out sparrows and vermin, sunk at least 150mm (6in) into the ground with a further 30cm (12in) bent out flat at right angles underground to prevent rats and other mammals digging underneath.

Should you be building a number of aviaries side by side it's worth boarding the intervening walls in addition, to prevent the cocks from seeing one another. The roof should be screened by nylon netting of a similar mesh size to the wire, to prevent the birds from hitting their heads should they fly straight up after having been frightened.

The shelter attached to the aviary should be draught-proof, with as much light as possible. Perches of 50 x 50mm (2 x 2in) timber should be placed quite high up in the shelter for your birds to roost on at night.

Feeding

During the winter the basic diet can be a 50:50 mixture of corn and pellets, either in a dish placed in a sheltered area, from which they can eat ad-lib, or provided twice a day on the ground, with the birds taking as much as they can eat during each session. Then, just prior to the breeding season (February to March), gradually change to breeders' pellets and continue with these until the birds have stopped laying (June to July).

Clean water must always be available. Titbits can be given in limited quantities to tame your birds, and a little fruit – such as halved apple and pears – can be given from time to time.

Partridges

These game birds are normally grown only by commercial organisations, for release for shooting.

Pigeons

These can be kept for meat, exhibition, or racing. You'll need a specialised structure in which to house them and keep them safe from predators and bad weather. This is generally called a loft or coop and is normally constructed to be both pleasing to the eye and functional.

There are three main types of pigeon, each one with a dedicated following of fanciers:

Above: Pink Breasted Racing Pigeon.

FLYING OR SPORTING PIGEONS

These are pigeons kept and bred for their aerial performance. They include racing pigeons and homing pigeons, which have been used to carry messages during times of war.

FANCY PIGEONS

Birds specially bred to perpetuate particular features, including breeds such as Jacobins, Fantails, and Pouters. Like chickens they have a breed standard that's applied when competing in exhibitions.

UTILITY PIGEONS

These are bred for their meat and as replacement breeding stock. The meat of pigeons is customarily referred to as squab and is considered a delicacy in many parts of the world. An example of a utility variety is the Kings.

Ostriches, emus, and rheas

All of these are mainly kept for commercial purposes, and specialist knowledge is essential.

ESSENTIAL INFORMATION

Daily care

Chickens are interesting animals. They appear to have absolutely no need for anyone to care for them and yet if trained will come running to you for the tastiest titbit or their daily handful of pellets. In the wild they would have ranged far and wide looking for feed, but we keep them in an artificial environment, and as such need to make sure that their daily needs, both physical and emotional, are catered for. Typically the novice keeper will find much of the day will pass simply watching their chickens, and this is no bad thing in itself, as it familiarises you with your birds and you begin to recognise patterns of behaviour and understand what needs to be done every day and why. A typical day will normally start when you rise for work or begin your normal routine, and with a little initial effort you'll soon be able accommodate the needs of your chickens into your own busy life.

6:00am (or when you rise)

This is the easy part of the day, when the birds can be let out from their nightly incarceration. Any cockerels will have been crowing since just before first light (although if disturbed by nocturnal visitors, be they mites or foxes, they will crow even earlier), and the hens will be bustling about getting ready for the first eggs of the day.

Above: The dominant hen will usually emerge first.

Below: Tomatoes are a welcome treat.

Above: Feed hoppers should be refreshed daily.

This is a crucial time of day if you want to discourage egg eating, as the morning disturbance as the birds move about the coop often results in early eggs suffering broken shells, as the limited space means the birds interact with each other much more and squabbles can break out, so try not to put off opening time until later.

Before the chickens emerge, the feed hoppers – which should have been left empty overnight to deter vermin – should be filled (see page 66), and the drinkers scrubbed and filled with fresh water. Make a mental note of any houses that need cleaning as you collect the early eggs, and always wait for all of the birds to emerge, normally starting with the dominant hen, who'll shake and fluff out her feathers, thus removing any sawdust and debris.

This is your opportunity to spot birds in poor health – which tend to be last out and very slow – and those coming into lay, as the cockerel will be hot on their heels. This is a good reason for deciding against having an automatic pop-hole (door) opener, which, although it would be invaluable to someone who has trouble getting up early, would prevent you from witnessing this important process. Watch out for potential problems such as a hunched stance (which could signify poor health), bullying, or feather pecking. Then walk around with your birds, talking to them as you do so which will put them at ease and make them that much easier to catch if there's an issue.

With the outdoor birds released, you can concentrate on any indoor stock – the exhibition growers that need protection from rain and sun (which can fade colour and ruin feathers), and those under-the-weather specimens in need of quarantine and a bit of TLC. As with the outside birds, feeders should be topped up and drinkers scrubbed. Always run your finger around the inside of the drinker – if it feels slimy, it will need cleaning. It should anyway be cleaned at least once a day.

Daily tasks

- Let the chickens out in the morning and check for their health.
- Wash drinkers and refresh water.
- Wipe and refill feed hoppers.
- Check grit hoppers and top up if necessary.
- Check for eggs after letting the birds out.
- Remove any droppings in the nest box.
- Refresh greens for indoor birds.
- Check on any sick birds and administer medicines as necessary.
- Check fencing for weak areas or holes.
- Feed grain in the afternoon.
- Spend half an hour watching for unusual behaviour or bullying, and enjoy your birds.
- Collect eggs regularly throughout the day.
- Lock up the birds at dusk.
- Empty the drinkers.
- Empty the feed hoppers, if they're accessible to wild birds and vermin.

Above: Check emerging birds for health.

Below: Rats will chew through wooden housing.

Weekly tasks

- Check the house and nest box litter and change or top up as necessary.
- Wash feeders.
- Check for lice and mites on the birds and in the house, and powder/spray when required.
- Keep an eye out for broodies.
- Check for rats and other vermin and take precautions as necessary.
- Check stocks of bedding, first aid equipment, delousing powders etc.
- Check the fit of leg rings (growers only).
- Check nails, spurs and beaks and trim if needed.
- Swap out non-performing males as necessary.
- Keep an eye on growers and move to bigger accommodation as they grow.
- Identify any birds to be eaten or taken to market.
- Carry out routine housing maintenance.

10:00am

You can now think about cleaning any houses and nest boxes. Although this is the most arduous and thankless of all of chicken-keeping tasks, the resulting muck is fantastic when dumped straight on to the resting bed of the vegetable garden to rot down for next season's crops. Make sure that any accumulations of droppings are lifted from the floor, and scrape the perches too. For more detail on cleaning, see page 64.

Cleaning also provides another opportunity to check the

Below: Wood shavings sold specifically for animal bedding is an excellent house litter.

birds' health. And if you carry a notepad and pencil around with you (avoid using a ballpoint, which will stop writing if it rains) you can make a note of any tasks that need doing and will be able to mark any fertile eggs so that you can record which pen they've come from. Keep your eye open for broken runs, damaged housing, rat tracks and parasite infestations.

This is usually the time of day at which you can remove non-performing males to let one of the newer and more vigorous cocks take over for a while, although you may find it easier to introduce him later, at sundown, in case the more aggressive breed females decide to attack him.

If a particular pen isn't producing fertile eggs, check that bums are clipped on the fluffier breeds, and also that claws and spurs aren't too long (and beaks too while you're at it), and trim them if necessary. Powder any birds that are showing signs of lice infestation.

1:00pm

If you're at home, check for any eggs and top up drinkers if necessary.

3:00pm

Time for another egg run, to catch fertile eggs before they're sat on; you'll also discourage egg eating, or even catch the culprit. You can scatter some whole wheat (one handful per bird) on the ground as you check the nest boxes and top up any drinkers that have run dry. Finally you can carry out any quick jobs such as oiling hinges or clipping wings before shutting the birds up at dark, which will depend on the time of year.

Below: You should record your daily observations.

Seasonal care

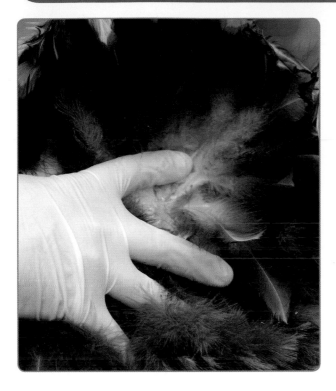

Above: Feathers should be parted to check for lice at the base.

- Barrier Red Mite Powder – dusted on to the birds.
- Barrier Red Mite Concentrate – diluted and sprayed into the house with a knapsack sprayer.
- Poultry Shield – diluted and sprayed into the house with a knapsack sprayer (but also available as a ready to use spray).
- Mite-Kill Spray – an aerosol, sprayed into the house.
- Diatom Powder – dusted round inside the house.

Above: Fill the sprayer with water, and then add the chemical.

Below: Cleaning should not be a chore.

As well as regular daily jobs there are also seasonal jobs that will periodically need incorporating into your routine.

Check your birds regularly to make sure they're not infected with lice or mites, which can spread disease and reduce the egg-laying capacity of your hens. This is done simply by parting the feathers under the wing and around the vent and keeping an eye out for any parasites that scuttle back under cover and for the presence of any nits (louse eggs).

Traditionally you could be relatively sure that the proliferation of certain parasites was controlled by the seasons. Typically a keeper would check for worms in the spring, lice and mites in the summer, worms in the autumn, and crest mites in the winter. More recently, however, and especially with warmer winters and wetter summers, it makes sense for poultry keepers to be continually vigilant for parasites and to incorporate checks for them into their weekly routine.

Check too for red mite, under perches and in crevices – felt roofs and the joints in tongue-and-groove woodwork provide a secure haven for these pests. Duramitex is an excellent treatment, but is now being withdrawn as it contains malathion; if you use it you should pay particular attention to the instructions and follow the recommended precautions. Alternatives include:

Above: A lick of preservative will convert a rabbit hutch into an isolation coop.

TOP TIP

A rabbit hutch makes an ideal house in which to isolate or quarantine birds.

Hens are more aggressive in the spring as they come into lay and seek to establish a new pecking order, with the dominant female having the best choice of nesting places. However, aggression can actually occur at any time when the balance of the flock is upset by the addition or removal of a bird or birds. As has already been mentioned, cuts on injured birds can encourage further pecking, so always keep a bottle of gentian violet spray handy, and be prepared to set up a rescue house should the need arise and one of the birds is attacked badly.

Below: A broody will sit tight in the nest box.

Keep an eye out for broody hens as you collect the eggs – they're the ones that huddle with fluffed-up feathers and scream when you try to pick them up. Be wary of a well-aimed beak, although it's more likely to surprise you than to hurt. If the hen is seriously sitting she may have pulled out her breast feathers to make better skin-contact with the eggs. Your choice is either to let her carry on or try to break the habit.

If you wish to use her as a broody, or egg-incubating bird, you should remove her to an isolated house and follow the process described on page 107. Then, once you're sure of her intention to sit, you can place some fertile eggs under her. Alternatively, if you want her to end her broodiness and resume laying as soon as possible you need to remove her to an anti-broody coop to break her of the habit (see the section on broodiness in section 5).

Egg eating can be discouraged by blacking out the nest box as much as possible with a black bin bag cut into ribbons and secured over the inside entrance, and by adding crock eggs which produce nothing when pecked, so that the bird tires of the exercise.

The longer days of summer encourage the birds to wander further away from their houses, and they may elect to lay in the oddest of places, so keep an eye out for stray eggs, which will attract winged predators such as crows and magpies. Similarly they may decide it's a good idea to perch in a tree or hedge, so you may want to clip the first ten feathers off *one* wing to unbalance them. Continue with your regular vermin control, as rats and mice can make a meal of feed bags; if you haven't already done so, you should get a plastic or metal feed bin.

Above: A black bag cut into ribbons will darken a nest box

Introduce fresh vegetable matter into your chickens' diet in the form of greens from the kitchen or weeds from the garden. Indoor birds will benefit from a feed of fresh vegetation, which can provide both exercise and entertainment if hung just above head-height so that the birds have to jump for it.

Should you wish to increase egg production in the winter the option of providing extra lighting is a matter of personal preference, though the consensus view is to let the hens rest over the winter to recharge their batteries. Since hens are born with a finite amount of egg cells in their bodies if you increase the number of eggs laid you'll decrease the number of years that they'll lay for. It also makes sense to let the hens' insides rest also, as too much forced laying can increase the chances of prolapse and shell or egg deformities.

However, if you decide to operate a yearly restocking system you'll find it useful to increase the light to 14–17 hours daily by means of a simple light (as an easy rule, one watt per bird, so 60 birds will need a 60W bulb) and a timer, whereby the extra light is provided in the morning and lets the hens go to bed with the natural dusk. This should be provided after the hens have experienced a period of short days, otherwise you'll not get the best results. Many people allow the birds to experience normal-length days up until November, and then increase the light from December onwards.

Although you may not notice any signs of worm infestation, control with a recognised product such as Flubenvet or Verm-X should be carried out in spring and summer.

Seasonal tasks

- Order leg rings.
- Carry out show preparation.
- Arrange breeding.
- Black-out nest boxes when laying commences.
- Review and update housing; buy or make additional units.
- Clip vent feathers of fluffy breeds to improve fertility (if you're not showing).
- Worm birds in spring and autumn.
- Move housing to new grass if the existing area is denuded of vegetation by scratching.
- Check hens aren't laying elsewhere, especially in summer.
- Check and clip wing feathers after the moult.
- Ensure the new feather stubs of moulting birds aren't being pecked.
- Adjust lighting levels according to season, to promote laying.
- Carry out house and fencing repairs, and repaint to ensure continued weatherproofing.
- Provide protection from the elements according to season – shade in summer, wind or rain protection in autumn and winter.
- Apply petroleum jelly to large combs and wattles in winter.
- Review stocking densities – hatch or buy-in new birds and dispose of unwanted ones.
- Adjust feeding regimes depending on age, health, breeding condition, exhibition status, or according to what's seasonally available (eg tomatoes in late summer and early autumn, root vegetables in winter).
- De-ice drinkers and clear snow if it falls during the winter.

Below: Crock (also termed china) eggs

A chicken-keeping diary

January

Midwinter and the start of the year, when you can begin your breeding plans to ensure healthy birds for the year to come. Pens or cages of breeding birds should be put together with second or third year birds; this ensures longevity in the line and will also give you an idea of how their progeny fared from the previous year.

Depending on the breed, five hens to one cockerel is about the right ratio to ensure maximum fertility, and the male should be introduced to the pen about three weeks before the eggs are required for incubating, to allow adequate fertilisation. He will need some time to settle in and you should watch out for any injury, which is the biggest problem, as the hens may not be too receptive at first. You may well notice sniffles and sneezes too, as a result of the stress caused by the introduction, so keep an eye on things and take action if necessary. It would be a good idea, too, to get your incubation equipment out and test it to make sure everything is OK.

Ensure that the hen house is as warm as possible, free from draughts, and fully stocked according to the manufacturer's instructions, as the hens will generate their own heat and too few birds will mean they get too cold and disease will proliferate. Any frostbitten combs should be treated with an antiseptic spray, plus a little petroleum jelly

rubbed in as a precaution on nights when the temperature will drop. The cold can also freeze eggs, so make sure that any destined for the incubator are collected regularly.

Some breeds of pullets (for example Silkies) will be laying, and those that are a little shy in producing can be encouraged by feeding them a high protein diet and providing extra day length by means of a timer.

Feed a good balanced layers' ration in general, and you can vary the afternoon grain feed by giving wheat, oats, barley or maize – alternately or as a mixed ration – as a change from whole wheat.

Traditionally, this was a time to cull out or sell unwanted birds that had been retained 'just in case'.

Below: A trio (one male and two females) is a typical breeding group.　**Above:** The spikes of the comb are easily damaged by freezing temperatures.

February

The end of winter at last, and as the snowdrops begin to emerge you will see more action as the hens get more outdoor time. Having said that, make sure you keep an eye out for the inevitable snow showers and plan your clearance operations in advance. Make sure that you have a snow shovel on hand and a working kettle to defrost the drinkers; better still, empty them at night and refill with fresh water in the morning. Nothing will stress the hens more than the sight of snow, and they may not come out at all, as the ground will have changed colour from what they're used to. Similarly, if the ground is white for a while and the birds have become accustomed to it, a sudden thaw will promote the same behaviour due to the sudden change to green or brown of what was once white!

Continue to feed a good layers' ration and keep up the supply of grain, interspersed if possible with vegetables such as turnips and swedes, but in small amounts to ensure that the birds aren't alarmed by the change. Also, ensure that the vegetables are removed if not eaten, so as to discourage vermin and to prevent frost ruining them, as frosted vegetable matter should not be fed to chickens.

The eggs from the breeding pens (the breeding flocks) should be fertile by now and you can set them under an experienced broody hen or placed in the incubator to check for fertility. If setting under a broody, do not put more than seven to nine eggs under her, as exposure to the cold may affect their viability. You could also, once fertility has been tested, consider selling them as hatching eggs. Thoroughly clean and disinfect any incubation equipment prior to and post incubation.

Below: Ensure snow is cleared from houses.

> ### TOP TIP
>
> To test fertility, run the incubator for ten days loaded with eggs and then candle them. Those that are fertile will have telltale starfish-like veins showing. You can then either dispose of the eggs or hatch them.

Above: Even ex battery hens will benefit from February sunshine.

Any chicks hatched in February can be allowed out at every available opportunity when it's not too cold, to harden them and to give them access to the vitamins available from the grass (which should be cut short) and sunshine. Care should be taken to prevent them getting wet, by providing a covered shelter close by to afford protection in an emergency.

Traditionally, February was considered the best time to start breeding for table poultry.

March

March winds are a sure sign that the warmer weather is approaching, and the hens certainly appreciate the extra sunlight, with more and more coming into lay. Fertility will also increase and money can be made by supplying hatching eggs, as the price will be greater than that for eating eggs. March is a very busy month between looking after the chickens, continuous hatching and the disposal of surplus eggs.

Fertile eggs that go by post require very careful packing and special polystyrene boxes can be bought for this. Great care should be taken to see that the eggs are packed tight so that the contents won't be unduly shaken; if the preformed hole in the packing is too big you can use

> ### Fact...
>
> Farmers traditionally fed their birds swedes because they believed it increased the size of their eggs.

Above: Your packing should be as professional as possible.

kitchen roll or bubble wrap round the egg to give a tighter fit. It is recommended that you place the polystyrene boxes inside a double-skin cardboard box for greater protection before sealing with tape and labelling.

If the weather is mild let any chicks outdoors as much as possible so that they can benefit from the grass and sun, but ensure that there's always someone around to keep an eye on them, as a sudden storm of hail, rain or snow can prove fatal.

Early spring rain will soon make the ground sodden, so you should put plenty of clean straw, coarse bark chippings or gravel down around the houses, which will keep the hens out of the mud and stop filth accumulating on the eggs.

Traditionally, March was the time to keep broodies and incubators in full production and move from hatching light breeds to heavier ones.

Below: Sudden snow falls will make the entrance to houses muddy.

April

The middle of spring, and time for unexpected showers of both rain and snow, so make sure that you have a supply of straw handy if the ground becomes waterlogged, and make sure any chicks that have been moved outside have access to shelter.

Chicks hatched in April will thrive, and this is probably the most popular time for non-exhibitors to breed. Keep a watch on the chickens for parasites as the weather brightens, and make the most of the warmer weather by thoroughly disinfecting your poultry houses. You'll now be able to separate some of the cockerels hatched in January from the pullets, as their glossier ornamental feathers will be well developed. With the birds being allowed greater free-range and access to plenty of insect life and fresh grass, you'll notice a reduction in feed costs if you ensure that you don't leave excess laying around for wild birds to pick over.

Traditionally, day old chicks were sold for high prices at this time of year, as people unable to breed their own replacements wanted stock.

Above: The back feathers of a male will be narrow, pointed and glossy.

May

Build or purchase an anti-broody coop if you need a task to keep you occupied. Hens that started laying earlier in the year, especially the heavier breeds, will be increasingly prone to becoming broody, and if you're collecting eggs to sell then this is not what you want. Ensure that the bottom of your anti-broody coop is made of heavy-gauge wire or strips of wood, thus making it uncomfortable and discouraging the broody by circulating cold air. Keep sorting out cockerels and ensure that young chicks get all the food they can eat by providing a covered hopper that can be kept permanently filled. You can continue to hatch, but if you do so make sure that you get the eggs set as soon as possible. Traditionally late spring was the time to fatten table birds for selling.

Above: A good anti broody coop should have an open base to allow circulation of air.

June

The increasing temperature means that you need to turn your attention to keeping the birds cool and their feed and water palatable. Early summer means that shade will be required for all chickens and if you have no natural shelter

Below: Straw bales provide shelter from sun and wind.

you need to provide some form of cover – even hurdles or straw bales will provide adequate shade, as will a sheet stretched over a line. By allowing your birds out of their coops as early as possible they'll benefit from the cooler temperature and make the most of foraging before the midday sun forces them into shadier areas. Keep their drinking water out of the sun, and refresh it at least three times a day to stop the growth of algae, which can prove fatal. Continue to look out for red mite in all the crevices of the coop.

Traditionally poultry housing was lime-washed or creosoted at this time of year to kill insect pests.

July

The chickens are more or less fending for themselves by now, and you'll find spare time to do odd jobs such as repairs or painting.

Food and water should continue to be provided – water especially so, as the hens are in full lay and need a constant supply. Keep an eye on your growers and if you notice any that are stunted or not growing as quickly as the others you should

remove them and house them separately, in order to ensure that they're getting enough food and don't suffer from bullying.

Don't keep unwanted cockerels, as hard as it may be not to, as they'll soon eat into any profits, and the older they are the more difficult it will be. If you can bring yourself to do it, learn how to dispatch them from an experienced keeper.

Traditionally, July was a month for repair of equipment and the initial assessing of stock for selling.

August

Late summer, and any growers will by now be approaching adulthood. Continue to provide extra feed to any slow-growing youngsters – there's nothing better for these than adding pet food such as wet cat food or dry biscuits.

Below: The brilliant red comb of a hen in lay.

Above: Odd jobs can be caught up with.

Birds that hatched early may be so advanced that their combs begin to redden in advance of laying, something that you want to postpone as long as possible to ensure that they lay through the winter. By keeping them on hard food (grain) for a time you should be able to ensure that no pullets are in lay before the end of September.

The moult may well now be starting in your more prolific laying hens and by feeding them only hard grain for about three days you can accelerate feather fall. Once the moult is in full swing you can begin to feed them properly again.

Traditionally, all cockerels needed for showing or breeding the following year were put in their cockerel pens and any others disposed of.

Below: The paler comb of a pullet approaching lay.

Above: a moulting hen will become pale and will stop laying.

September

Move your young pullets into the laying houses where they're to commence laying. This ensures that they get over the shock of moving and come into lay by the end of the month or a little later. Birds that come into lay now should, with only the occasional pause, continue to lay all through the winter, when the older hens are resting and will need plenty to eat. The older birds will still be moulting and you'll see a drop in egg production from them as a result. Don't forget to spray your houses regularly, and be vigilant against red mite, which can proliferate seemingly overnight.

This is also really your last chance to get any repairs done, as early autumn heralds quite strong winds that can soon damage houses in even slight disrepair.

Traditionally, due to careful planning, pullets and hens overlapped in their laying seasons in order to guarantee a continuous supply of eggs throughout the year.

October

Keep up to date with your house maintenance (including cleaning) and repairs this month, to ensure that all the coops are weatherproof before the worst of the weather sets in during mid-autumn. Although dry cold doesn't do any harm to the birds (as long as you grease large combs and wattles), and will be of benefit in that pests are destroyed, damp and draughts are the greatest cause of illness to your hens and you should be alert to these conditions. To keep their houses and litter bone dry, try scattering corn on to the litter to get the hens to scratch round and keep it friable.

As nights and mornings are colder, it's advisable to see that birds are allowed access to food all day, as no insects are about and this shortfall in nutrition must be supplemented with a good layers' ration.

Traditionally egg prices rose at this time of year, and those who delayed hatching at the beginning of the year by as little as a fortnight lost a month in egg production at this time of year.

November

Late autumn, and a good month to take time to reflect on your hatching plans for next year.

Mixed corn can be fed every other day as an alternative to whole wheat, while oats are also acceptable during the cold weather to provide internal warmth. The first of the nationwide shows, the National, is held this month and exhibitors will be making their final selections and getting their birds ready for showing.

December

Well-kept pullets will be approaching full lay by now, and some of the older hens – particularly Silkies – will have started laying again. Give the birds extra grain at night, as they'll be locked in their houses for many hours during early winter and will need the extra to get them through the dark hours. Keep an eye out for any cracks and openings in the house, including the floor, and make them draughtproof. Chickens will generate enough heat (provided the house is correctly stocked) to keep themselves warm. Don't allow them out of doors in bitterly cold winds or rain, unless you have suitable protection in place, such as bales set out in a cross pattern to provide shelter from all directions.

Below: Straw bales laid out in a cross pattern provide shelter from the elements in all directions.

Cleaning

Above: Use dust-free wood shavings.

Though cleaning is a chore to some, to others it's a chance to spend time with their birds, to check their health and to delight in the amusing antics they get up to, from dust-bathing to squabbling over a tasty morsel.

When to clean is a subject of great debate amongst keepers, with some cleaning daily while others think twice a year is more than sufficient. However, it's basically more or less determined by such factors as the weather, whether the birds roam and, of course, how many birds are kept together.

In reality a pristine house is not a good idea. You should avoid being too clean, since as long as the birds remain healthy exposure to a little muck does them good, because it helps them to build up a basic immunity to disease, whereas if the house is too clean there's no challenge to their immune systems and they'll fall victim to any disease that's going around. After a while you'll be able to recognise when your house needs cleaning, although if you can smell ammonia or the shavings are damp then it's already well overdue. Damp litter should be avoided at all costs, as many pathogens thrive in it.

What you should use is the next question. A good choice is wood shavings (buy the pet-bedding quality, dust-free type) or cross-shredded paper for the inside and straw for the nest boxes, in a layer at least 25mm (1in) deep. Avoid hay or long-shredded paper, which can tangle round birds' legs and often cakes with their droppings and festers, encouraging toxic mould spores; and don't use finely chipped bark, which can also contain spores. If you want to use a deep litter system for economy of time and money then straw is probably the best medium. Coarse bark or pebbles in the run outside are both fine, and help to keep the chickens' feet off wet mud.

The process of cleaning is simple. Just remove all the muck and shavings with a small shovel (a hand-held coal shovel is ideal) and a rubber trug, or take out the droppings board if there is one, and either dump it all on your compost heap or tie it securely in plastic sacks if you don't want the bother of a heap, where it will happily rot down to an excellent compost for next year. You can use a simple rotation system here and throw the shavings straight on to a resting vegetable bed

Below: Bark containing cocoa or showing evidence of mould should never be used.

The deep litter system

A deep litter system is one in which a layer of straw is placed on the floor of the henhouse to a depth of 15cm (6in). As it becomes dirty, more straw is added, in effect making it deeper. Then, once or twice a year, the whole house is emptied, disinfected and refilled. Typically this method of housing is used in enclosed, commercial systems or large breeding establishments, which are de-stocked when the time comes to clean them out (ie the birds are either slaughtered or moved to alternative accommodation).

ready to dig in during the autumn. However, putting the muck straight on to a growing bed as a mulch isn't recommended, as the dung has a high nitrogen content and will burn foliage. Disinfect the house with a proprietary poultry disinfectant, paying particular attention to the corners.

Remember too that drinkers, feeders and grit hoppers will need regular cleaning. First you need to ensure that the contents are removed. Then, using warm water to which a mild first aid disinfectant has been added, along with a squirt of washing-up liquid, wash the outside and inside with a scrubbing brush. Don't use Jeyes Fluid or other substances containing phenol, as this is poisonous to poultry.

Feed hoppers can be wiped out with a damp cloth weekly unless fouled by droppings, although they should be washed as above at least once a month to ensure cleanliness. Drinkers, however, should be properly cleaned every day, as you can't see the contamination that may be in them. You should scrub the insides thoroughly to remove any trace of algae, especially in warmer weather, when it can grow daily.

Once washed, you should rinse the containers with clean water to remove all traces of the cleaning solution. Then dry them thoroughly before refilling, especially the feeders, in which food may otherwise stick to any residual water and become mouldy. Don't refill drinkers from water butts, which may have been fouled by other animals or contain poisonous algae or other diseases.

Grit hoppers benefit from a monthly emptying and cleaning to remove any stagnant water or germs trapped inside by the grit. Also, check underneath all containers for signs of red mites, which will hide anywhere that is dark.

Above: Clean the brooder thoroughly between hatches.

Perches should be scraped and disinfected every two days to two weeks – depending again on the density of birds being kept. You should clean out nest boxes once every two months. Other equipment such as incubators and brooders will need cleaning before introducing another batch of eggs or chicks.

Get yourself into a regular routine, working through your daily, weekly and monthly tasks (see the suggested checklists on pages 53 and 54), and there's every chance that very soon what began as a series of chores will become second nature and you'll find you actually start to enjoy it.

Below: A hunched hen indicates a problem, as in the case of this ex battery layer.

Feeding and watering

What exactly do you need to feed your birds to get best results? Everyone's advice will differ, but you need to do whatever is suitable to your own situation and budget, and is obviously best for the birds.

In the wild

Free-range chickens are naturally omnivores, picking their way through greens, grains and whatever tasty piece of protein blunders their way, and this same attitude can be seen reflected in our garden birds, who relish everything we put out for them from nettles to cat food, and even mice and frogs should they be unfortunate enough to get into the run. However, if you're keeping your birds in a closed or restricted house they can't access this natural larder and it consequently falls to you, their keeper, to provide them with a sensible and balanced diet.

The process

A chicken's beak is much more delicate than you might think, being more sensitive to texture than taste. Just watch the response of a cockerel when you throw a titbit into the run – his first job is to pick it up and drop it before calling the

Above: The beak is a very delicate tool.

others over. He has literally 'felt' if it's good to eat. This is one reason why pellets resemble grains in size and shape, and are a preferable option to mash, which is best used to stop anti-social behaviour such as feather pecking by keeping the hens engrossed in picking up the tiny pieces.

Once swallowed (unchewed), food passes to the bird's crop, which bulges and distends often to quite alarming

Below: Birds will free range naturally.

proportions, depending on how much the bird has taken in. Here it is softened by lubricants before passing to the stomach (proventriculus) for chemical digestion and the gizzard for mechanical digestion, where small ingested stones grind it. This is why indoor birds benefit from flint grit (as opposed to oyster shell grit, which provides calcium for bones and egg shell quality) from the age of about ten weeks, to allow the gizzard to develop properly.

In order to make best use of this process, feed your birds pellets during the day (about two and a half to three handfuls per bird) and whole wheat in the evening (about one handful per bird), as the latter is digested much slower and will get the bird through the night's fasting. Remember also that too much of a good thing will fill the crop (birds are naturally greedy) and result in an imbalance of intake and consequently nutrition.

How much a bird will actually eat will depend on its sex, age, and general state of health, as well as its breed, the season, and the extent of its activity during the day.

Types of feed

Whether you're keeping a few birds for eggs or table, for exhibition or just for fun, one thing you'll have to work out is what to feed them. Traditionally special concoctions were made up according to secret recipes, jealously guarded and handed down from father to son and never let outside the family. Nowadays, however, compound feeds are specially formulated to provide birds with all their essential nutrients and vitamins, and can be supplied broken up as chick crumb, powdered in the case of mash, or with a higher protein content for growing stock in the form of growers' pellets.

The ration may contain coccidiostats (ACS) to provide youngsters with a basic immunity against coccidiosis (an intestinal disease – see section 5), so check the label depending on your preference, as you may wish to foster

natural immunity. Yolk enhancers aren't necessary in feed, as grass will naturally deepen the yolk colour, and antibiotic content should be avoided.

Check if the ration is labelled 'natural', since there's currently no legal definition of what this means and genetically modified maize and soya may be included. Your conscience therefore needs to guide you here, and if you're in any doubt go for a feed labelled 'organic' instead, as these are legally defined and will contain no artificially extracted soya (since the process by which it's extracted has been reported to use carcinogenic chemicals) or genetically modified ingredients.

Complete feeds

Commercially available feeds come in both mash and pellet form, both containing exactly the same nutrients but with pellets being a compressed version. The advantage of mash is that as it consists of smaller pieces the birds literally have to work for their supper, having to invest more effort into picking up enough to make a decent meal. This in turn prevents them getting fat and reducing the number of eggs they lay and reduces boredom, which is the first step towards many nasty vices. It can be fed dry or wet, although if wet you must remember to remove uneaten excess from the run before it spoils, while if fed dry the same advice still applies, as it does tend to scatter everywhere, enticing in vermin and often fouling drinking water.

Pellets have the advantage of looking like grains or tiny worms, and so are seen to be the more natural option. They're also easier to handle and ensure that the birds can't simply pick out the best bits as they would do with a mixed feed.

If you have crested or bearded birds then pellets must be your first choice, as mash can get stuck in damp feathers and before you know it the other chickens will be investigating this mobile food source and will be pecking out feathers, leading at best to a ruined topknot and at worst to a bout of cannibalism.

Below: The elements of an essential diet – grit, whole wheat and pellets.

Below: Pellets.

Above: Whole wheat.

Grains and seeds

These can be purchased either as mixed corn or whole wheat. Mixed corn will often contain wheat, oats, barley and maize, and is great to use as a treat, since the birds pick out their favourite morsels, invariably the maize. Use whole wheat if you can get it to prevent this selecting behaviour, and feed it as an addition to the pellets/mash rather than instead of, to prevent the birds filling up on it and getting fewer nutrients. Ideally you should feed it to them a couple of hours before bedtime, so that it sits in the crop overnight and digests slowly, keeping the bird going as it slumbers. Oats can be used in the winter as they're a good source of slow-digesting carbohydrate during colder periods.

Buying and storing

Having researched and decided on your preferred feed, be it mash or pellets, organic or not, try to get to a busy specialist supplier, such as a feed merchant, who stocks or can buy-in a number of different brands and types. The busier the supplier, the more likely they are to be reliable, as it means that people are using them regularly and their stock is turned around faster, not left to fester in a corner.

Check that the dates on the bags are still current and not likely to expire before you use up the contents, as nutrients do break down over time, becoming either ineffective or – worse still – toxic. Also make sure that the bags are free from tears and water marks. If you can, ask to see a sample of the feed. Pellets should be firm and any grains shiny – never take a bag with musty or shrivelled seeds.

Make sure you have a suitable vehicle available to

Above: An open feed sack will attract rodents.

Below: A rodent proof feed bin is essential to keep feed from spoiling.

Above: Flint Grit.

Above: A flower pot makes an ideal grit holder.

Below: A wire peg inserted through the drain holes will secure it in the ground.

get the feed home, as the bags are normally very heavy (around 20kg) and you may want more than one. Be sure to buy only as much feed as you can use within six months, as nutritional content will reduce over time as a result of exposure to heat, light and air. As soon as you open a bag, tip the contents into a firmly closed, vermin-proof bin, and don't top up with fresh until the entire old bag has been used up.

Keep the bin in a cool, dry place out of direct sunlight, and keep it closed to stop mice getting in along with dust and other debris, and to prevent acceleration of the breakdown of those precious vitamins. Mice will habitually urinate as they go, contaminating grain and feed, which can lead in turn to intestinal problems in the birds after they've consumed it.

Grit

Though specialist feeds will contain calcium, as it's needed for both bone and shell formation, you can supplement this if necessary with oyster shell grit should egg quality be poor.

While oyster shell is optional, flint grit is a necessity and should be permanently available to your birds, as it helps to grind food in the gizzard in much the same way as millstones grind grain into flour. In growers it even assists in the formation of this organ.

Although both types of grit are available separately they're usually sold as 'mixed' grit by pet shops and feed merchants, due to consumer demand and cost. However, don't be worried by this, as all you need to do is fill a hopper with the mixture and the chickens will pick through it and only select the pieces that they need. Some fanciers will mix a little pet-shop charcoal in with the grit too, to aid the digestive process and keep the crop and gizzard healthy.

The best dispenser for grit is a flowerpot into which a bent piece of wire can be threaded to secure it to the floor. The holes in the bottom of the pot allow rainwater to drain through.

There have been suggestions recently that providing oyster shell grit adversely affects the absorption of other minerals, but unless you're seeing an increase in problems associated with bone formation or egg laying you should continue to allow free access to both types.

A deficit of calcium can affect pullets coming into lay quite severely, as they undergo considerable stress as their bodies adapt to producing eggs. A hen will need to supply approximately 2.4g (0.085oz) of calcium daily to the oviduct to make her egg shells, and if sufficient calcium isn't available from her normal diet she'll use calcium from her skeleton. The bones then become weakened to such an extent that birds are unable to stand, the sternum and ribs become deformed and bones are easily broken. This is often diagnosed as cage layer fatigue.

Above: Chickweed was traditionally used as chick feed.

Greens

Access to vegetation is equally important, whether in a run or hung up in the pen, as the birds will digest the cellulose from it in a separate part of the gut, absorbing the nutritional element and voiding the residue as that sticky, foamy mass you see about every tenth dropping.

Below: Windfall apples will be eaten ravenously.

Make sure that any vegetable matter is fresh and free from pesticides and other pollutants (don't pick it from the roadside, for instance). If your birds are restricted indoors, hang the plants in the run to provide amusement and keep the birds active. Any uneaten matter is best removed at the end of the day to prevent rotting and to keep vermin at bay. Suitable vegetable matter includes:

- Berries – raspberries, blackberries, strawberries.
- Greens – broccoli, cabbage, kale, spinach.
- Weeds – plantain, nettles, chickweed, parsley, Fat Hen, Good King Henry, dandelion, Shepherd's Purse.
- Fruit – apples and pears.

Protein

Chickens enjoy animal protein every now and again, and will normally pick up bugs and worms as they scratch around, even devouring larger prey such as mice if they can catch them; so if you keep your birds indoors, or if their run is limited, you may want to feed them something like mealworms, which are available from pet stores (sold as reptile feed). However, moderation is key since mealworms, like maize, contain a lot of fat, and you don't want to encourage the birds to put on extra weight. Mealworms are best kept cool and in the dark to slow their development, and should be brought back up to room temperature before feeding to make them more active.

Protein is used in the chicken's body to produce, amongst other things, feathers and antibodies. During times of

feather replacement (*ie* the moult) or illness your birds' protein requirement will increase, and unless you adjust their diet accordingly health and resistance will suffer. If you're feeding them pet food, note that cat food – both wet and dry – tends to be a better alternative than dog food, as it normally has a higher protein content.

A bird that moults continually or has brittle feathers is likely to have a protein deficiency or be suffering from an illness, possibly digestive, that is inhibiting protein absorption. Excess protein in the diet can also cause problems, for example gout.

Scraps

Legally, if you're selling your eggs you can't feed your birds any animal material or derivatives; but your average chicken, kept for purely household use, will love the odd titbit, though again you must make sure that they don't overindulge, as a fat hen won't lay well or live that long. Boiled potatoes (the forerunner of commercial chicken 'mash'), fresh kitchen scraps, scrambled eggs, and crushed and baked eggshells are all popular choices, as are excess courgettes, cherry tomatoes and strawberries from the kitchen garden. Cooked meat should be avoided, as it will encourage rats, and anything rotten is not a good idea as it has health implications.

Feeders

The design of feeders can have a direct impact on the health of your birds, as droppings can accumulate in them and if left on the floor pests can seek shelter underneath. By fitting a guard to the top you can prevent the birds perching on them, while suspending them from the ceiling will stop birds scratching in them. If your chickens develop the habit of tossing the feed out of the hopper, try replacing it with one that has a lip that curls inwards, or try raising the whole thing to just above the birds' back height.

Drinking water

Keep it fresh, keep it clean, and keep it available at all times, as a chicken's body is 50% water, while an egg is 65%, meaning that this is probably the most important part of the daily diet. Unfortunately a chicken is programmed to drink wherever it sees water, more often than not from a dirty puddle which can harbour any manner of diseases, so ensure that the run or pen can't accumulate standing water by improving drainage. If it's on a solid base, ensure it slopes away from the chicken house.

The average laying bird will drink up to 500ml of water a day, by taking a sip and then tipping her head back to swallow. She will drink little and often, increasing in

Above: Feeders and waterers come in many different forms.

regularly if suffering from nutritional deficiency or illness, but will drink less if she's not laying. If a hen is deprived of water for 24 hours, it may take up to 24 hours to recover, during which time it's unlikely that she'll lay. If the period is over 36 hours, she may go into moult followed by a period of poor laying.

Below: Any water source, no matter how dirty, will be used by your birds.

Above: Drinkers come in many different shapes and sizes.

Water needs per 12 birds per day

Age	Requirement
1–7 days	1 litre
1–4 weeks	2 litres
4–12 weeks	4 litres
12 weeks plus	6 litres

Use a suitable drinker such as the plastic ones specifically sold for the purpose, and ensure that the bottom lip is wide enough for the chicken's head to reach the water, as different breeds will need slightly different widths. Also, note that new chickens brought in – especially if you have ex-battery birds – may not be used to that particular style of drinker and will need careful watching to ensure that they do drink. If you think they're having problems you can temporarily use a deep dish.

Below: Drinkers are best emptied during the winter to prevent freezing which may split the container.

Top up your drinkers from a tap, *never* from a water butt, which can harbour diseases associated with parasites and algae.

Once a week, add one tablespoon of baking soda per gallon (4.5 litres) to automatic drinkers to stop the build-up of slime. Clean and disinfect them regularly. In winter ensure that drinkers are de-iced, and in summer keep them in the shade out of direct sunlight.

If the drinker is of the galvanised type, ensure that no rust is present, as this may indicate damage to the zinc coating, which can lead to poisoning.

If you have a large number of birds and find the container constantly needs refilling, a larger drinker might be more suitable, as these have a reservoir that refills the receptacle if it drops below a certain level. Try to avoid open bowls if you can as these easily attract wild birds, which will drink and bathe in them, again potentially spreading disease.

Nutrition

Free-range birds are able to forage and will therefore get a lot of the essential nutrients that they need from their surroundings, including sunlight, and this, supplemented with a complete ration, will give you a healthy bird able to go about its daily business and resist disease. Some nutrients are manufactured inside the chicken's body, so factors affecting this process, including stress, will have a major impact on its overall health.

Below: Situating the drinker on a bed of pebbles will prevent standing water.

Above: A large drinker will have a self filling reservoir

Below: Supplements should be added according to the manufacturer's instructions.

Vitamins

Vitamins have their origins in plant and animal matter, and are essential to the diet. They're of two types: those that are retained in the body fat of the chicken, namely Vitamins A, D, E and K; and those that are used as needed and the excess expelled, mainly C and the B complexes, which as such need to be ingested more regularly.

Chickens are most likely to be deficient in A, D and B2, with birds raised in cages being more prone to a lack of vitamins as they can't pick through house litter to get access to some of them. Housed birds in turn will have less access to the correct intake than free-range birds. As already mentioned, stress is a major cause of deficiency as ultimately the chicken eats less and therefore reduces its amount of vitamin intake; similarly some diseases can reduce appetite and/or reduce the efficiency of synthesis within the body.

Vitamin	Source	Used for
A	Cod liver oil	Vision, bone growth, disease resistance
B and Complexes	Grains, leafy greens,	Growth and hatching meat, brewer's yeast
C	Fruit and vegetables (chickens make their own Vitamin C so supplements are not necessary unless stressed)	Reduces stress
D	Cod liver oil, sunshine	Strong bones, beaks, claws and shells
E	Wheat germ oil	Fertility
K	Greens, spinach	Blood clotting

A good feed ration, access to free-range and sunlight should give your birds all the vitamins and minerals they need for a normal healthy life, although a supplement that can be added to their drinking water is useful during stressful times, such as when moulting or unwell.

Minerals

These give bones and shell their strength and are used by the body to interact with other nutrients. As with vitamins, caged birds are more likely to suffer a deficiency than housed birds, which are in turn more prone than free-range birds.

Mineral	Source	Used for
Calcium & phosphorus	Oyster shell grit	Bone and eggshells
Magnesium	Feed ration	Bone and eggshells
Potassium	Feed ration	Egg production
Manganese	Feed ration	Bone and eggshells
Copper	Feed ration	Colour
Selenium	Grains	Fertility

Housing

Above: An onduline roof should be sloped from front to back.

For large-scale breeders, an abundance of poultry housing in all sorts of shapes and sizes means that they only need to give their old coops a good clean with a proprietary disinfectant before they introduce new birds. However, for new keepers – and, indeed, even for some of those who've kept poultry for some time – the availability of so many types of housing makes it hard to make the right decision when it comes to buying a new chicken house, with options ranging from flat-packed kits to those that can be delivered already assembled or even erected by the manufacturer.

However, before choosing a house you should bear in mind a number of factors:

- Material – pressure-treated timber is the best choice if you decide on a wooden house, although a number of plastic-based materials are also used.
- Roof – needs to have a slope and be able to channel water away from the pop-hole. Onduline is popular, although felt is cheaper. Red mite will find felt easier to hide under, but onduline can also be affected.
- Ease of movement – needs to be borne in mind, especially if you're limited by space and need to make frequent moves to fresh ground, since you don't want to be lugging a hefty house around in the winter when you get home from work! Many houses can be fitted with wheels.
- Ease of access for the keeper – you need to be able to clean the house properly and collect eggs or catch birds. A few bangs on the head will soon make you realise if the house is impractical.
- Ease of access for chickens – the pop-hole (the doorway through which a chicken can pop in or out of its house) should be lockable to keep predators out and to keep the hens confined at night.

- Aesthetics – ideally a house should be pleasing to look at, especially if it's to be situated in the garden. Wooden houses can easily be painted a different colour if you change your mind.
- Position of nest boxes – do you prefer internal or external? Either way they must be dark and higher than the floor and provide a snug area in which the birds can lay. Check if the lid on external boxes is waterproof, as this is a common fault in houses made by people without a working knowledge of poultry.
- Ventilation – this is very important. Too draughty or too stuffy and your birds will become prone to disease.
- Size – really does matter. You may well start with three hens, but consider if you're likely to get another three. However, don't make the mistake of too much space thinking you're being kind, as a few hens in a larger house can't maintain their body temperature in the winter and will soon become ill. If the manufacturer says it's a five-hen house, then it's just that, not a seven or a three.

Below: The house should be easily opened for cleaning.

Above: The nest box lid should be suitably waterproof.

Below: Sheds are easily converted into poultry houses.

Omlet

Above: The Eglu.

It's always best if you can to visit manufacturers or sellers beforehand to get an idea of how their houses perform. Alternatively get their catalogues and brochures and do a bit of research yourself. Internet forums are very useful when it comes to people giving you their opinions. Three of the most popular suppliers are Omlet, Littleacres and Forsham Cottage Arks, although there are many others out there equally deserving and a little Internet research will soon get you their details.

Omlet (www.omlet.co.uk) supply the Eglu, a popular choice for the new keeper since it combines many good features and comes in five vibrant colours. Fitted throughout with wooden roosting bars and an integral nesting box, the chickens are kept warm in the winter and

cool in the summer due to its twin-wall construction. Collecting eggs is made easy by a convenient side door. Cleaning is simple thanks to a slide-out droppings tray and fully removable lid. Made from energy efficient material, it's claimed that the Eglu will last for years and at the end of its life can be 100% recycled.

Both Littleacres (www.littleacre-direct.co.uk) and Forsham (www.forshamcottagearks.com) are traditional-style houses favoured by the more serious fancier and breeder and are always represented at the major shows and exhibitions – which is ideal if you want to 'try before you buy'. Their products are long-lasting and can cater for a larger number of birds and have a wider range of styles.

None of the housing options are cheap, but you do

Above: A Littleacres house is an attractive addition to any garden.

Above: Battery hens emerging.

get what you pay for and the after-sales service is superb. If you're not happy these suppliers will go out of their way to ensure that the customer is satisfied.

Once the house is installed, all you need to do is put the birds in gently, preferably at night as they'll be stressed from their journey and very nervous. So handle them gently and talk to them quietly. Then confine them in the house for 24 hours with a feeder and drinker, which will ensure that they know where they can get security and quiet and will encourage them to go in at night. The following morning, you just open the pop-hole and let your chickens come out in their own time. If you've sourced ex-battery hens, then this is a critical time and it can take many days before they emerge; but have patience and you'll be rewarded when they take their first hesitant steps into their brave new world.

One last thing – consider adding a protective run. It will protect your birds from dogs and foxes and even over-zealous children, who can cause them alarm if the birds aren't used to them. At the same time it will confine them to an area where damage can be limited, or where ground clearance can be maximised depending on your point of view.

Above: New hens can be encouraged out the following morning.

Below: A run will protect your beds and keep them restricted to a particular area.

Budget house-building

If you're more of a DIY keeper, and cost is an issue, you may like to try building your own house, where you have control over the materials (for example, you may like to recycle).

What you'll need

- External-grade plyboard (4mm/0.16in is used here).
- Lengths of sawn wood (25mm x 25mm/1in x 1in is used here).
- Wood screws.
- Hinges and turn-button for the door (or extra wood if you want to make a sliding pop-hole).
- 1cm (0.5in) gauge wire for the ventilation hole.
- Power drill.
- Jig saw.
- Electric screwdriver.
- Tape measure.
- Set square (to make sure edges are straight).
- A staple gun and staples (to secure the ventilation wire).
- Suitable roofing material.

Preperation

Cut the ply as follows:

- 1 @ 100cm x 100cm (39.37in x 39.37in) for the base.
- 1 @ 105cm x 105cm (41.34in x 41.34in) for the roof,
  to allow a slight overhang, which you can make bigger or smaller as preferred.
- 1 @ 75cm x 100cm (29.5in x 39.37in) for the front.
- 1 @ 65cm x 100cm (25.5in x 39.37in) for the back.
- 2 @ 75cm/29.5in (front edge) x 100cm/39.37in (width) x 65cm/25.5in (back edge), ie the edge meeting the front is 75cm (29.5in) and the edge meeting the back is 65cm (25.5in), to give you a slope from front to back when assembled.

Cut the lengths of sawn wood as follows:

- 2 @ 92cm (36.22in) for the front legs.
- 2 @ 83cm (32.68in) for the front cross-pieces top and bottom.
- 2 @ 82cm (32.28in) for the back legs.
- 2 @ 83cm (32.68in) for the back cross-pieces top and bottom.
- 2 @ 75cm (29.5in) for the front left and front right side uprights.
- 2 @ 65cm (25.5in) for back left and back right side uprights.
- 4 @ 92cm (36.22in), for the left and right side cross-pieces, top and bottom.
- 5 @ 100cm (39.37in) max – or depending on how you want the lid to fit – for the roof and a base cross-piece.

1a Make the front first, adding the pop-hole and a ventilation hole. Starting with the front – as this is the most intricate – decide and mark where you want the pop-hole and ventilation hole. Then, after making pilot holes with the drill, cut out the shapes.

Fit the wire over the ventilation hole and secure to the internal face with the staple gun, making safe any staples that protrude. The pop hole door can be fitted at any point, but it may be easier to do so when the house is assembled, so that you can judge the position of the hinges and turn button and check for correct alignment.

1b On the internal surface, ie the face that will eventually be inside the house, attach the legs, setting them in from the side edge by the thickness of the ply plus the thickness of the leg (in this case, they were set in 4mm/0.16in + 25mm/1in = 29mm/1.16in from the edge), and secure flush with the top edge. You'll find it invaluable to drill pilot holes in the wood before securing, in order to stop it splitting when the screw goes – you can buy a special drill bit to do this. The leg will then stick out past the bottom of the panel. Add the cross-pieces to the top and bottom of the panel, between the leg struts.

1c A sliding door can be added to run inside two pieces of grooved wood at a stage in the assembly where you are happy with its position.

2 You can now make up the back in the same way, remembering that it will be 10cm shorter than the front to allow for the slope, but the legs will protrude by the same amount.

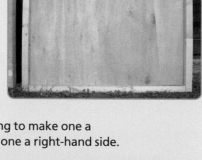

3 The sides are just as easy, although there's no need to inset the uprights from the edge – you simply frame each side piece with the sawn wood, remembering to make one a left-hand side and one a right-hand side.

5 By measuring the internal dimensions at the base, a bottom is easily cut and screwed in, though you may find it useful to fit an additional crossbar under the floor for greater stability. Similarly, measure the internal dimensions across the top and make up a frame of 2.5cm x 2.5cm (1in x 1in) wood and secure it to the top panel, which will be slightly bigger than your frame, allowing it to fit snugly as a lid and overhang slightly.

4 Once all of the parts are made up you can get on with the task of assembling the unit. While you can do this on your own two people will make it easier, as one can hold while the other screws. Start with one side and the front – the side should butt up nicely to the inset legs – followed by the other side and the back.

6 You can then add Onduline for a better finish (if using this, remember to make sure that no nails or screws point through to the living space), or felt if you prefer, which is quicker and can be secured with galvanised nails.

If you wish to fit a nest box, make sure it's higher than the floor, as naturally a bird will nest off the ground to avoid predators and rising water levels, while an internal perch needs to be higher than the nest box, as a bird will always roost at the highest point. To make the nest box, you'll need to assemble a box consisting of three sides and a base along with a lid. You then attach the open side of the box to the side of the house, in which you'll have cut a hole through which the hens can enter.

A coat of a good preservative will treat the bare wood and make for longer-lived accommodation – and with the many different shades available, you can stain it whatever colour you like. Many good brands will also be 'bat friendly', meaning that they're also suitable for poultry. Leave it to dry for the recommended time to ensure good penetration and to allow any vapours to disperse, and then you can add or return the chickens.

Alternative styles

Once you've mastered how the house fits together, why not try using different materials, such as tongue-and-groove timber, and try changing dimensions for different birds, such as youngsters etc. If you're interested in tongue-and-groove, it's better to make up panels first by attaching the wood to the frame before cutting the boards to fit, rather than trying to fit the cut-to-size pieces directly to a frame.

Above: You can experiment with different designs.

Above: Try experimenting with tongue-and-groove.

Above: A tongue-and-groove side panel with integral nest box.

Above: Even this anti-broody coop is based on the same design.

Security and biosecurity

Above: Even reliable pets should be excluded from the chicken runs.

Too many times is the tale of woe heard from both new keepers and experienced poultry people (who should know better) regarding mayhem in the henhouse resulting from a predator attack. You thinking that no fox could get over a six-foot wall and that your birds are much happier roaming about in the garden rather than being kept in a run does little to guarantee their safety, as predators actually come in all shapes and sizes. So regardless of the fact that you may

have never seen a fox, or your confidence that your dog wouldn't hurt a fly, or any other reason for not securing your chickens, always make sure that you take appropriate steps to keep your birds safe.

Electric fencing

This option is probably the best all-round security for your chickens. It can be as complicated or as simple as you wish, and can take the form of either single wires or netting. Power is supplied either from a battery or mains – if the latter, you'll need to consult a qualified electrician for the installation. As electric fences of the wire-strand type are designed to keep larger livestock within a boundary they won't keep foxes out, so you'd need additional precautions such as an internal wire fence. Electrified poultry netting is therefore far more suitable.

A basic set-up will consist of a net, an energiser, a battery and an earth stake. The net surrounds your chickens, house and all, and is ideal to keep larger breeds and hybrids from wandering, as well as keeping unwanted visitors out. Once in place the energiser, which converts the battery or mains power (depending on your setup) to a high voltage pulse of

Below: Internal fences will provide further protection as well as segregating stock.

Above: Birds learn to respect electric netting.

electricity, is connected to the net with one cable, to the earth stake with another, and to the battery with the two terminal cables. A leisure battery suitable for caravans is the best option rather than a car battery, as it's designed to be on constantly.

The final part of the circuit is supplied by the animal itself when it comes into contact with the net – it grounds the pulse through its feet and the shock is felt. Depending on the energiser settings, you may be able to increase the pulse in order to train the birds to respect the net, which means that as soon as they touch the electrified wires with their heads the current is grounded through their feet and the shock results. It may be distressing for you at first, and certainly the chickens get a shock from the power, but they usually only touch it once and then keep their distance.

It's advisable to be in attendance when you first set it up, as a frightened bird may run forward, trapping itself in the wires, in which case you'll need to turn off the energiser and release it. A really determined bird will run straight through the netting in a bid to escape the electric shock, so again you'll need to be on hand to catch it.

Fox-proof fencing

A fox-proof fence should consist of a thick wire mesh at least 2-3m (6ft 6in) high, secured to concrete posts and buried to a depth of 60cm (24in), with the leading edge angled out underground away from the enclosure so that if the fox digs down, he'll hit wire both in front of him and below. At the top there should be an additional section about 30cm (12in)

wide and again angled at 45° away from the enclosure, to prevent a determined fox from climbing over.

A wooden fence is a less secure alternative, but in many gardens it's the only option. You can still try to ensure that it's as secure as it can be by adding strips of angled wire as described above, buried in the ground at the bottom and attached along the top.

Biosecurity

The following biosecurity advice is given by the Poultry Club of Great Britain:

- Keep feed under cover to minimise wild bird attraction.
- Keep water fresh and free of droppings.
- Keep waterfowl and chickens separate.
- Control vermin.
- Quarantine new stock for two to three weeks.
- Quarantine birds for seven days after taking them to an exhibition.
- Change clothes and wash boots before and after visiting other breeders.
- Change clothes and wash boots before and after attending a sale.
- Keep fresh disinfectant at the entrance to poultry areas for dipping footwear.
- Disinfect crates before and after use, especially if lent to others. However, it's preferable not to share equipment at all.
- Disinfect vehicles that have been on poultry premises, and avoid taking your vehicles on to other premises.
- Comply with any import/export regulations/guidelines.

Predators

Humans

Unfortunately, in this day and age theft is an all too common problem, especially with some exhibition lines being very expensive or hard to get hold of, and also due to the continuing interest in cockfighting. Keepers' chickens are stolen (or worse) by human predators, who take birds to order for unscrupulous breeders or for baiting fighting cocks, or to create mayhem just on a whim. You should always lock your chickens up at night as a matter of course, and you may like to add a padlock to the house if you think theft may be a problem.

Battery-powered shed alarms, which are triggered when a door is opened, are invaluable if your birds are to be kept in any form of outhouse, and if you have a house alarm you can even get alarm system add-ons which will cover your sheds. Along with CCTV cameras, which are becoming ever smaller and available as wireless versions, a multitude of security options are there if you need them.

Any form of deterrent must be worth the effort. Even a basic security light situated near the coop, which will be triggered by an intruder, makes a useful alarm.

Having ensured that all housing and outbuildings are as secure as you can make them, remember to check the integrity of the buildings, as thieves may find it easier to break into the non-alarmed and unsecured back of a house through rotted wood or via holes that can be prized open. Also, try not to keep all your good stock in the same area, so that if theft does occur you won't lose all of your line breeders. Look at your boundaries, and take the same

Above: CCTV is now compact and affordable.

precautions as you would to secure your own property. Spiky hedges such as hawthorn and berberis are a superb deterrent, and an added trellis panel on top of fencing makes it harder to climb over. If you're particularly isolated, why not set up a guineafowl cage for the benefit of their alarm calls?

As a further precaution, if you sell birds or eggs don't give your address freely – it's better to give your telephone number and get to know any potential customers before arranging to let them on to your property. That way, at least you'll have some form of contact details should the unthinkable happen. Try not to give visitors tours of your stock areas either, however eager they may be, as you may be inadvertently giving a potential thief a good look round

Below: Small alarm systems will protect your outbuildings and alert you to intruders.

Below: Be wary of requests for all black or all white birds.

at what you have to offer. It's better to have the birds boxed ready for them in another part of your property.

It is also necessary to warn you about people interested in your cockerels if you have a quantity for sale, as there's a fair demand for bait birds, which are used to train fighting breeds. Unless you're prepared to accept this risk, try to enquire what the birds are wanted for, and don't be afraid to say no if you have any doubts. Be wary too of people just wanting pure white or pure black birds with no real interest in the breed, as these may be destined for ritual use – not a nice way for a bird to meet its end.

Birds and animals

Vermin

While there is no strict definition of the term 'vermin', poultry keepers would consider winged vermin to include crows, rooks, magpies, jays, sparrows, pigeons and starlings, while ground vermin would include rats, mice, weasels, stoats, mink, foxes and grey squirrels. The fact that you have a concentration of poultry in a relatively small area will create a natural draw to vermin, as they'll seek out the associated foodstuff, offspring and even the birds themselves.

As with many problems, prevention is always better than cure. By investing in proper storage containers for feed you'll limit the amount of grain and other food that's accessible to rats, mice and sparrows, and you should combine this with an organised method of feeding

> **Fact...**
>
> Rats dislike the smell of mint, and rat-catchers of old would soak rags in mint to discourage them.

to further limit the available food – for instance, by using covered feed hoppers in runs.

Security of your flock must always be a top priority, as nothing can decimate years of hard work quicker than some of the more vicious vermin. Poultry are a prey species and are always at their most vulnerable at night, when they roost, so limit the access that other animals may have by providing good fencing and secure housing.

If you do find you have a vermin problem, then a control method should be used in conjunction with your prevention measures. Although there are live trap options, 'control' usually refers to killing, and includes trapping, poisoning and shooting, although all methods will need a degree of expertise. If in any doubt at all you should consult your local council, who will be able to put you in touch with pest control services.

One option available to the more humanely minded is the growing range of sonic deterrents, which are triggered by movement and emit a high pulse of sound waves that cause discomfort to the animal that sets them off. Specific types can be found for scaring cats, birds, moles and even foxes.

Below: Close gauge welded mesh is the best option to keep wild animals out.

Foxes

Despite the old wives tales that abound, foxes don't kill for pleasure, they kill to eat, and they can wipe out an entire coop in a single night if you let them. Their main method of killing is to break the neck of the bird with a shake of the head, and although death is usually quick, if disturbed a fox may drop the bird before the fatal injury is sustained, leaving you with a dead or crippled chicken or, if you're lucky, with one that's merely shaken (though it may later die from the shock). Each body is then carted off to a secure place and buried before the fox returns for the next corpse.

Lazy when full, a fox will always seek an easy meal, so if your chickens are locked up but those of your neighbour allowed to roost outside, then it will be your neighbour's birds who'll suffer. A determined fox can manage to scale a 2m (6.6ft) fence if hungry enough – hence the angled top to a fox-proof fence (see page 83).

The main times of activity for foxes are January, February and March, when their cubs are in need of feeding, and again in June, July and August, when the cubs become more independent; but they're still active at other times, so constant vigilance is necessary. Control, where you have an ongoing problem, is best undertaken by an expert, since shooting, snaring or trapping can cause undue suffering if carried out incorrectly, and while you don't want to see your chickens killed it's illegal to cause suffering to any wild animal.

Above: Run a triple line of electric wire around the perimeter for security.

Fact...

A fox's eyes will reflect pink if a light is shone at them at night.

Older and more traditional methods of deterring a fox include allowing your dog to walk around the chicken enclosure, as he'll almost certainly mark his territory as he does so, which in turn will deter a fox. Human urine is also a deterrent, although, as you'd expect, it's not an option favoured by everyone! There's also a fox-repelling oil, although it shouldn't be used to excess as the fox will then become used to the smell.

One of the best methods of deterring foxes has got to be an electric fence, which will give the inquisitive nose of the intruder a short, sharp shock and is otherwise harmless. Don't be afraid that children, cats and other household pets will be harmed, as, like the fox, they get just a small shock that will keep them away but will cause no further injury. However, do keep an eye out for hedgehogs if you install electric netting, as they curl into a ball once shocked and invariably curl around the wire. The preferred method would be to run a triple wire around a fenced enclosure, high enough to allow hedgehogs to crawl under it but low enough to still stop foxes.

Rats and mice

Second only to foxes, these are the animals which poultry keepers typically consider to be vermin, and they're a particular nuisance when you keep chickens. It's not the chickens that attract rodents, but their feed, and bad feed storage and feeding practice quickly leads to a bad rat or

Above: Rats will try to get to any chicken feed.

Above: A cat is a useful part of your vermin control.

mouse problem. Both will climb walls with ease and will happily chew through housing, making them a costly pest to eliminate.

A good outdoor cat (Siamese crosses are the best) will keep the mouse population down, while a trained dog (a terrier is best) will track down and destroy nests of rats, and normally this is enough when combined with good husbandry. However, if you have a serious problem you may well consider traps or poisons, both of which are effective. If you're using a dog, make sure it can't get access to the runs, as many will become indiscriminate in their killing once they start and you'll lose the very thing you're trying to protect.

Traps are of the 'jaws' type, which are designed to snap shut and kill the animal outright, but they still require a little skill in use to be effective. They should be installed in a tunnel for best results, as most vermin hate being exposed and will stick to sheltered runways. You can make use of hedges, walls and ditches, and if setting a trap in a man-made tunnel make sure that the materials you use are local to the area in order to blend it in with its surroundings. The construction should be wide enough for the trap to shut correctly and to allow you to maintain it.

Above: Keep poison tunnels topped up.

Should you find that a rat is still alive in the trap, a strong swift blow across the nose or a pellet from an air rifle should kill it instantly. Remember to wear gloves to protect yourself from Weil's disease, which can be fatal.

Poisons are an alternative option if you have a large population, or if the animals have become wary of traps, although they're distressingly indiscriminate in their effects, as they'll kill most things that eat them; and as they have an anti-coagulant effect on the blood, the animal bleeds to death, which is neither quick nor painless. A simple bait box or tunnel is a good choice, as other animals can't get to the poison and you'll just need to remember to change brand every month or so in order to prevent resistance to a particular chemical.

Mice will also take rat poison, but quickly become wary of it, and if affected can die in the open, making them an attractive food option to other animals, which become poisoned in turn.

Below: Rodent trap.

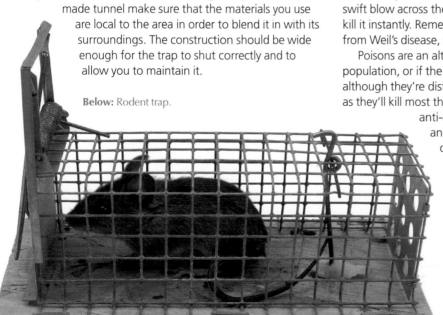

Above: Mink.

Above: Crows can become a nuisance in poultry runs.

Mink, stoats and weasels

These members of the weasel family are all ruthless killers of poultry, often leaving no trail, as they kill swiftly and drag the carcass away immediately. Their high metabolic rate makes them aggressive and constantly active hunters, able to scale fences and squeeze through the smallest of gaps in search of their quarry, with young birds being especially vulnerable.

Notoriously difficult to keep out of a poultry run, they do find smooth metal such as galvanised steel difficult to scale, but you may find it unsightly to surround your runs with this, in which case trapping with something like a fen trap is recommended. Make sure you follow the instructions carefully when setting these, as they can be dangerous if handled incorrectly. If you find any rat holes or mole runs make sure you fill them in, as weasels and their ilk will use them to gain access to your birds.

Magpies and crows

Most of the crow family are black or grey, or pied in the case of magpies. A variety of crow species breed in the UK and can be a particular nuisance, as they're highly intelligent and soon learn where you put out feed for your chickens, waiting for you to vacate the area before they move in. Jays, magpies, rooks, ravens and jackdaws are all members of the crow family and all will take eggs and youngsters if they can, as well as adding to your feed bill.

Controlling crows with a gun is difficult, as they're alert to the colour of the human face and take to the air as soon as you look at them to shoot. However, you can use this to your advantage by floating a pink balloon in a chicken run if you have a large problem. A more reliable option for control is a Larsen trap, which is set up with a live decoy bird in one half. The live bird is enticed in with offal or a nest made of straw

> ### Fact...
> A sudden decline in the local moorhen population may indicate the presence of mink.

and set with a couple of hen's eggs, and once caught is allowed to hop around in a closed area of the trap. Crows won't tolerate another bird in their territory and are drawn to the hopping captive, and in an attempt to engage it and drive it off are caught in the other side.

The captured bird can be released away from your property but may be attacked by others if you introduce it to the established territory of other crows, so a more humane option might be to destroy it with an airgun if you're experienced (in order to prevent unnecessary suffering). If you're in any doubt about your capabilities on this score, or if you don't want to destroy the bird at all, contact your local environmental health department or wildlife centre for advice.

Types of crow

- The carrion crow is one of the most adaptable of our native birds and often the one that plagues chicken runs. Quite fearless, it can nevertheless be wary of man and is fairly solitary. It's very intelligent and has been known to leave nuts on the road for passing cars to crack.
- Rooks are easily distinguished by their pale, bare faces, while a thinner, light-coloured beak and domed head differentiate them from the carrion crow. They're sociable birds, congregating in noisy flocks which can overwhelm poultry.
- Magpies are black and white and will 'chatter' incessantly, especially if guarding territory. They are expert scavengers and predators and have been associated with the theft of eggs and chicks from breeding pens.
- Jackdaws are distinguished by the shiny cap on their heads and are usually seen in pairs. They're more of a nuisance in that they'll eat the chicken feed.
- Jays are the most colourful members of the crow family, although as woodland birds they're rarely seen. Their distinct alarm call, often copied on television murder mysteries, will let you know where they are. They're not really a problem for garden poultry.

Above: Unsightly but effective, a dead crow suspended in the runs will deter other crows.

> ## TOP TIP
>
> If you have a serious crow population problem, a dead crow hung up in the run will deter others.

■ The raven is a large, all-black bird (distinguishing it from the rook), with a large beak and long wings. A diamond shaped tail distinguishes it in flight. It has been a protected species since 1981, and isn't a real problem for poultry enthusiasts.

Pigeons and other wild birds

Although not an immediate threat to the lives of chickens and other poultry, crows, pigeons and starlings will eat their feed, so that it will cost you more to feed your birds, and they may carry diseases to which chickens are susceptible. The smaller and prettier birds – sparrows, finches, tits and others – aren't really a problem and can be lured away from the area by tastier options such as fat balls. However, larger birds such as blackbirds, starlings, doves and pigeons can be a real nuisance in poultry runs, eating feed and messing in the containers, and should be excluded as much as possible.

The only real solution is to totally enclose your chicken house with a run made of 1cm (0.4in) weld mesh and feed your stock with a good quality layers' pellet in galvanised or plastic containers. Whole wheat should be fed in the evening by scattering directly into the grass, to prevent it from lying around in feed hoppers. Fruit-cage netting is another alternative, and being temporary can easily be moved.

You can also try to deter wild birds by using some of the many gadgets available for that purpose. Polypropylene tape that hums in the wind is one sort of deterrent, but can become ineffective after about three days when the birds get used to the noise it produces. CDs spinning on canes are an alternative, as are 'owl eyes' plaques, which have a pair of eyes painted on them which spin in the wind – although on the downside, both also make chickens a little uneasy for a while and may temporarily put a halt to laying.

You can limit access to feed by purchasing feed dispensers that require the chickens to peck at a 'tap' before they're rewarded with a scattering of pellets. Though it may take less intelligent birds a little while to get the hang of it such feeders are effective, as not only are wild birds too small to reach the tap, but it also prevents their droppings from reaching the feed, so improving biosecurity.

Above: Humming line (also called buzz line) catches the wind and hums to deter wild birds.

Below: Owl Eyes spin in the wind.

Laying

There's nothing like the thrill of collecting fresh eggs from your own hens, and children especially get great pleasure from rummaging in the nest boxes to find them. Once collected, eggs are best stored in the refrigerator if you're going to keep them for any length of time, as this delays the ultimate breakdown of the contents. However, as the colder temperature causes the internal air sac to contract and microbes on the surface of the egg to be drawn inwards, always make sure that stored eggs are clean.

If you intend to eat the eggs within a few days you can store them in a cool room, as the bloom on the egg prevents microbes from entering.

The chicken and the egg

Even at birth, a female chick has a limited number of egg cells in her ovaries, and this is the maximum that – all being well – she will ever produce; and while we can influence how she lays by what we feed, how the bird is housed and how much light is available, remember that the more eggs a hen produces early on in her life, the less she will produce later.

Pullets (females under one year old) can start to lay as early as 16–20 weeks (traditionally 18) from hatching and are said to be at 'point of lay', *ie* about four weeks off of coming into lay. Imminent lay is heralded by their faces, combs and wattles becoming redder, and they may well start to make throaty noises often described as a sort of soft 'cawk cawk cawk' sound. The female will lay most of her eggs in her first season (which is why battery hens are often discarded in season two), following this with a moult of her feathers and growth of a new set. During this stressful process laying is suspended as the hen diverts all of her energy to replenishing her feathers. This period, incidentally, is usually quite a good indicator of how the bird is laying, as a good layer will drop all her feathers quickly and replace them rapidly, while a poor layer will drop only a few and regrow them slowly.

> **Fact...**
>
> Egg colour is normally related to the colour of the chicken's ears – red ears produce brown eggs while white ears produce white eggs.

Below: The fresh egg on the right sits higher, and the white spreads less.

Above: A wind egg (on the right) is much smaller.

During her second year the hen (a female over one year old) will lay fewer but larger eggs, so the overall produced weight over the two years remains about the same.

Colour of the egg is always subject to preference, with consumers preferring an earthy brown colour and its association with rural life, to pristine white and other colours. The blue/green of the araucana or legbar egg is simply the influence of blue pigment on a white or brown shell, producing china blue or olive respectively. Eggs acquire their colour in the oviduct of the hen, the last stage before laying, when pigment is deposited on the shell's outer layers. When you crack an egg you'll find the inside and the outside are different colours, with the exception of eggs laid by Araucanas, which are the same colour throughout.

Like colour, size and shape differ too. We are all familiar with the double-yolked egg, which results from two egg yolks being encapsulated at the same time, often when a bird is coming into regular lay and her system hasn't settled down sufficiently. You may come across small eggs that might not have a yolk at all – these are sometimes called cock eggs or wind eggs.

If you have a glut of eggs, apart from eating them or selling the surplus at the gate you'll find that freezing works quite well if you want to use them for pastries and cakes – just separate whites from yolks into ice-cube trays and freeze.

Selling eggs

An excess of eggs isn't in the same league as an excess of courgettes, as you'll always find someone willing to take them off your hands. Selling a few eggs direct to friends and family from your home is exempt from regulation – if you can't get your hands on a supply of egg boxes you can buy new ones cheaply in bulk on the Internet.

Regulations

If in doubt regarding selling eggs, it's always best to consult your local environmental health department or the Government website at http://www.DEFRA.gov.uk/foodrin/poultry/trade/marketregs/eggmarketregs.htm.

A point to note, however, is that since 1 January 2004 regulations have required all Class A eggs sold at retail level within the EU to be stamped with a code identifying where they were produced, the country of origin, and the method of production (ie organic, free-range, barn or cage). The aim of these regulations is to improve consumer information and choice, and to assist with traceability and enforcement of the EU Egg Marketing Regulations. However, since producers selling eggs to consumers directly from the gate of their home or farm have clear traceability regarding where the eggs have come from, there is no requirement to stamp them.

The Registration of Establishments Regulations came into force on 31 December 2003. These require all laying hen establishments to be registered and allocated a distinguishing number indicating the farming method, EU Member State code, and a unique identification number. Establishments with fewer than 350 laying hens and those rearing breeding laying hens didn't and still don't have to be registered under these Regulations. The benefits of such a registration system are the traceability of eggs, and information on the location of egg producers that can be used for other purposes, such as welfare and disease control.

From 1 July 2005 farmers selling eggs from their own farms at local public markets have also had to mark their eggs, irrespective of the number of laying hens kept. Some producers with fewer than 350 laying hens therefore had to register for the first time, in order to obtain a distinguishing number. Sales in farm shops and door to door, however, remain exempt.

Below: Araucana eggs are a delicate porcelain blue.

First aid kit essentials

Having got your basic equipment, and the birds housed, you need to give consideration to how you're going to keep them in good health. The basics of a workable first aid kit, which will contain all you should need, are as follows:

- Liquid paraffin – to ease an impacted crop, and for making crusty leg scales supple.
- Vaseline – for combs in winter.
- Haemorrhoid cream – reduces swollen prolapsed tissues, making treatment easier.
- Gentian violet spray – antiseptic and masks the colour red, making wounds less attractive to peckers.
- Vitamin and mineral supplement – helps sick birds recover.

- Probiotic – reintroduces gut flora.
- A calcium supplement to prevent soft-shelled eggs.
- Aloe vera nose and ear lotion – cleansing and antiseptic.
- Antibiotic eye ointment – obtainable from your vet.
- Surgical spirit – to treat scaly leg.
- Cotton wool and cotton buds – good for cleaning and applying lotions.
- Thin disposable gloves.
- A small soft toothbrush.
- Virkon S – a multi-purpose disinfectant (needs adding to water).

Below: Your first aid kit will consist of both prescription and non prescription items.

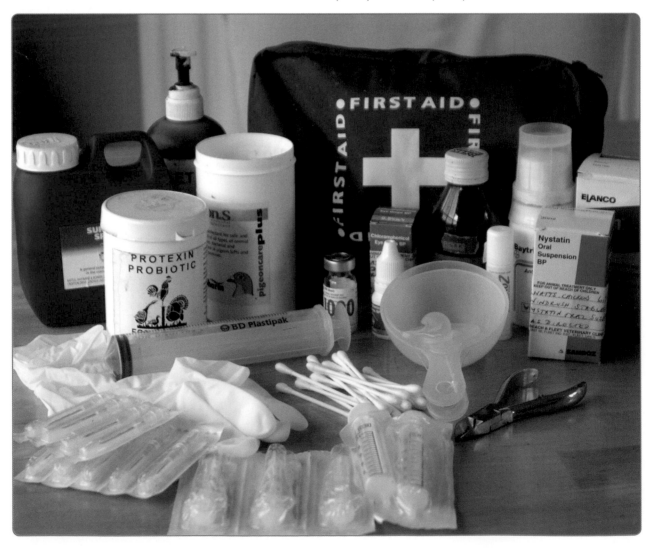

- Flubenvet or Verm-X – internal parasite treatment.
- Flea spray and louse powder – one for the house and one for the bird.
- Cider vinegar – acidifies the gut, making it more attractive to beneficial flora.
- Stockholm tar or similar anti-peck spray.
- Poultry saddle – for back wounds and mating damage on hens.
- Coccidiosis treatment.
- 21g needles.
- 1ml, 2.5ml and 5ml syringes, plus a larger one for administering fluids.
- Nail clippers – for beaks and toenails.

You can add to this as you go, but this list represents the most effective mix of products to start with.

An adequately stocked first aid kit should be considered to be an essential part of your chicken equipment. Some items will be used only in an emergency while others will be used on a regular basis such as louse powders and wormers, and in all cases gloves should be worn to prevent the transmission of disease both from and to the bird.

Not only will you be prepared for those emergencies which can occur at any time such as cuts and minor infections, but it will allow you to treat more life threatening conditions quickly, efficiently and without unnecessary and expensive trips to the vet. Experienced keepers can treat quite complicated conditions very effectively with just a few of the items listed in the recommended kit, while the novice will find they have the confidence to treat minor injuries straight away and so prevent an escalation of symptoms.

Learn as much as you can about anatomy and have a book such as Gail Damerow's *The Chicken Health Handbook* to hand both as a guide to what is ailing your chicken and to take to the vets should you need to.

Remember however that first aid is just that – it is the immediate aid you give to your chicken to prevent its condition getting worse; it should not be a substitute for veterinary advice, especially if you suspect one of the more contagious diseases, or if surgery is needed for a pet. In addition, many regulations exist regarding the treatment that you can give yourself and more specifically to drugs and their methods of administration, so if in doubt, consult your vet.

Above: Regular healthcare is as important as emergency measures.

Below: You should keep a variety of syringes in your first aid kit.

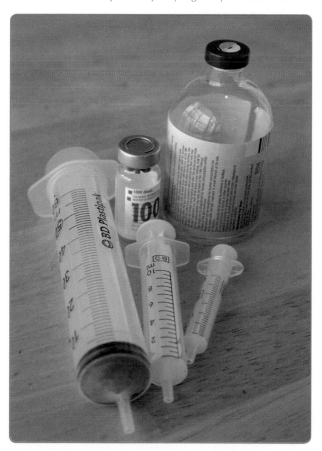

SECTION 4 Chicken Manual

SPECIALIST INFORMATION

Anatomy

One of the most recognisable characteristics of chickens is their constant scratching at the ground, which in the wild would enable them to unearth seeds and insects, along with small animals such as mice and lizards disturbed by their feet. As a consequence, the chicken's digestive system is designed to cope with all manner of food, and its good eyesight means that potential food is spotted quickly and targeted accurately.

Chickens aren't capable of sustained flight and will only take short flights to roost or to get a better view of imminent danger. In general the lighter the bird, the greater its propensity for flight, the larger breeds seldom achieving more than a half-hearted jump skywards. This is a result of their ancestors living on the forest floor under a dense canopy; the birds have evolved as ground-dwellers and their diet and behaviour have adapted accordingly. Even so, as birds, chickens have a very light skeletal system and a large muscle structure (the breast) which does allow some flight, and their feathers have become adapted to facilitate this as well as display and, to an extent, temperature control.

The forest-dwelling nature of the chicken's ancestors has also resulted in the need for vocal communication, in order for flock members to locate each other, their young or a mate. We're all familiar with the crow of a cockerel, and hens make noises too – calling their young to them (a task also shared by the male) and cackling when they lay an egg, in order to reinforce their bond with the flock (which may have wandered out of sight).

The normal lifespan of a chicken is five to eleven years, with hybrids living shorter lives than pure breeds as their bodies are designed to put all their energy into laying up to 300 eggs per year, as opposed to pure breeds which lay fewer eggs per year over more years. Not surprisingly, then, a lot of the chicken's anatomical and behavioural features are based on reproduction and display, with moulting and laying playing a major part in a bird's year.

Living in a large communal group termed a flock, a strict hierarchy exists to promote social harmony. There is a strict 'pecking order', with a dominant male and female having access to the best feeding and nesting areas. If this social balance is upset by removing a dominant bird or introducing another, fights will occur, sometimes resulting in serious injury, until the balance of order is restored.

As part of their communal lifestyle, hens will try to lay in nests that already contain eggs, and will often move eggs from neighbouring nests into their own, which is a practice utilised by breeders, who put crock (china) or rubber eggs into a nest to promote laying. Hens can be extremely possessive about particular nests, often even trying to share the same nest at the same time. The flock will therefore use only a few preferred nests, which makes locating their young and eggs harder for a predator but also means that once found an entire nest will be wiped out.

Below: Hens will share a nest.

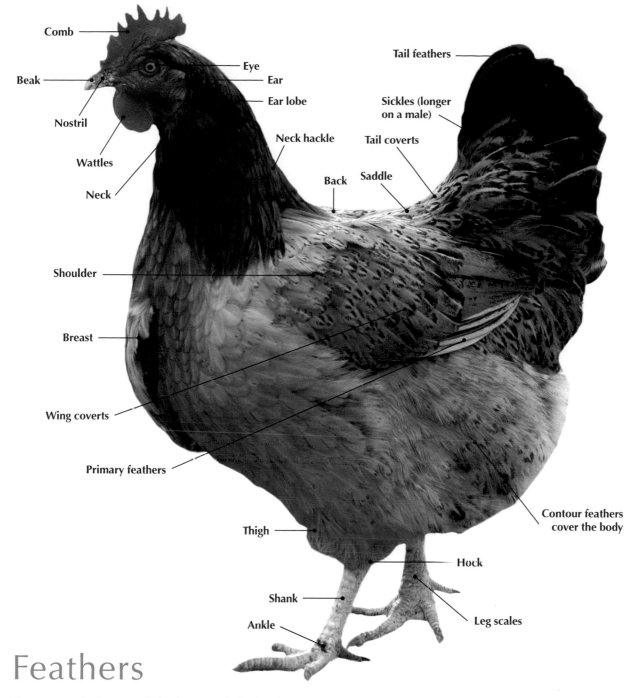

Comb

Beak

Eye

Ear

Ear lobe

Nostril

Wattles

Neck

Shoulder

Breast

Wing coverts

Primary feathers

Thigh

Shank

Ankle

Neck hackle

Back

Tail feathers

Sickles (longer on a male)

Tail coverts

Saddle

Contour feathers cover the body

Hock

Leg scales

Feathers

These are surely the most distinctive part of a bird and have evolved over time for the purposes of flight and display. In a chicken they are of four basic types:

- Contour feathers – are the most obvious and cover the wings, tail and body. They consist of fibres that interlock with the aid of barbs when the feather passes through the beak during preening, and with the addition of oil from the preen gland (situated at the base of the tail and looking like the tip of a biro pen) form a waterproof layer. When a new feather first emerges it's contained within a blood-filled sheath, which if broken can result in profuse bleeding. As the feather matures, the blood is withdrawn and the feather can be cut without any problem.

- Semiplumes – are the fluffy feathers that cover the body and act as insulation; they have a shaft like contour feathers.
- Filoplumes – are situated at the base of each contour feather and look like single hairs. They're especially noticeable on a plucked bird. These feathers are amply supplied with nerve-endings and help to keep the contour feathers in the best position; if you ruffle a bird's feathers it will instantly preen them back into position due to information received from the filoplumes.
- Down feathers – are present on chicks and to some extent on adult birds. They don't have a shaft.
- Feather colour is produced either by melanin (yellow, red brown, brown and black) or carotenoids from plants (yellow, red and orange).

Above: Saddle feathers are prominent in the cock bird.

Above: Feathers are replaced annually.

Cock birds

Males have slightly different feathers to females, and these are termed ornamental feathers. Found on the neck, shoulders, back, saddle tail and hackle, they're typically longer and shinier and have a softer texture. It's these feathers which can be used to start sexing birds at about eight weeks, as they'll be present in cockerels but not in pullets.

Head feathers

Crests (on the head), beards (under the beak), ear tufts and muffling (on the cheeks) all add to the appearance of the different breeds. The crest is also termed a top-knot or tuft and its presence usually means that the comb is small or placed at the front; the feathers form a rounded bonnet and in the male are ornamental, being longer and shinier. Beards result in smaller or non-existent wattles.

Below: The Poland's crest must surely be one of the largest.

The moult

Feathers are moulted and replaced annually, normally after the breeding season, and it requires a lot of the bird's energy to grow and replace them. The process will usually take between three and four weeks but can last for up to two months, during which time egg laying has to cease as the bird's reproductive system gets a complete rest. It's an extremely stressful time and often results in birds becoming lethargic; wattles shrink and pale out, and birds may even develop sniffles because of their weakened resistance to disease. Basically the chicken's body can't cope with regeneration and reproduction simultaneously.

Each new feather pushes out the old one from its follicle. This may happen over the entire body all at once, which leaves the bird looking like a hedgehog and very uncomfortable, or in a specific pattern, starting at the head (head and neck, breast and body, wing and tail). You'll notice that the new feathers are deeply coloured and shiny after the weathered appearance of the old ones. A good layer will drop all her feathers seemingly overnight and will recover quickly if in good health. A poor layer or an older bird whose laying capacity has diminished will lose a few feathers at a time.

The new feathers emerge enclosed in a blood-filled sheath, and it isn't wise to pick up the bird at this stage, as disturbing the new feathers may result in the sheath breaking and the new stub bleeding. The sheath is quickly shed as the feather matures and the blood supply is withdrawn, a process taking up to several weeks. Being made of protein, the feather replenishment takes a lot out of the bird nutritionally, so some keepers will add dog or cat food to their daily ration.

Males also undergo a moult. Though rarely as drastic as amongst the females, they'll more than likely be infertile while it's in progress, a condition that sometimes becomes permanent if the bird is underfed during this time.

Apart from the annual moult, young females (pullets) may undergo a neck moult when they've just come into lay, stopping egg production temporarily. This is caused by the body not being quite ready for the rigours of laying. It can be

Above: A good layer will drop all of her feathers in one go.

Above: The stress of the moult may enable illness to take hold.

prevented by keeping youngsters on growers' pellets until point of lay, when the wattles and comb begin to get larger and redder.

Chicks undergo a moult of their down feathers as their new feathers come through over the first couple of weeks of life, followed by an intermediate moult at about 12 weeks of age when their first-year adult feathers arrive. It's at this stage that their adult markings become more defined.

TOP TIP

When the moult starts to become noticeable in laying hens you can speed up feather fall without doing anything too drastic, by feeding just hard grain for three to four days out of every seven. Once the moult is in full swing begin to feed well again, and after a week or so you can put them back on to normal rations.

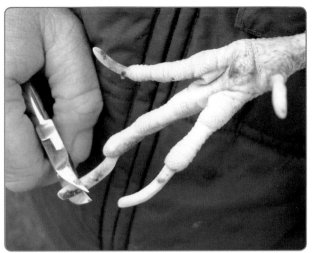

Above: Claws grow continually.

Claws and beak

Made from keratin, these will grow throughout the life of the bird. Both may need trimming if there's insufficient activity during the day to blunt them, but you'll need to pay close attention to the quick, as cutting into this will cause bleeding. (See also page 152.)

Spurs

These are present in both sexes, although they're much bigger in the male, and are used as a fighting aid to determine rank. Cockerels will have a rounded spur measuring usually less than 2.5cm (1in), with cocks having longer, sharper spurs that grow about 1cm (½in) every year, although older spurs are damaged easily and then splinter back to a shorter length. Older birds may have difficulty in walking or mating if their spurs are too long, and you should shorten them if this occurs. (See also page 152.)

Right: Spurs can be trimmed every other year.

Scales

These cover the legs and are shed once a year to be replaced with shiny new ones. They're kept in supple condition with oil from the preen gland.

Colours and markings

Probably the most attractive thing about the many different breeds of poultry is the diverse array of patterns and colours that can be found, from pure blue-whites to iridescent blacks that gleam with a green sheen when struck by the light. Chickens have certainly evolved to make the most of display and camouflage. There's a huge array of colours and markings, some breed-specific, some sex-related, and new ones are constantly appearing as the popularity of breeding increases.

Colour

Colour can be used to distinguish male from female and chicks from adults, and even gives an indication (in ducks) of the time of year, when the feathers become drab and the bird enters a phase called 'eclipse'.

Through centuries of mutation and selective breeding, it's the bird's genetic make-up which decides how the colours will display and what patterns will result. There are several oddities in the chicken world, for example the white-crested Poland, which has a bright white crest on an otherwise coloured body, and the white Silkie, which is basically

Above: Female partridge colouring is duller than that of the male.

Below: Silver laced is a black border on a white feather.

a coloured bird without a colour developer gene to 'turn on' the colour.

The original jungle fowl displays typical 'partridge' colouring, which in the male is a black head and tail (with a beetle-green sheen), orange/red saddle and hackles, red wing bows and back and orange breast, while the female is a dull brown all over with a paler, often salmon-coloured breast. This enables the cocks to be seen by potential mates in the undergrowth and hides the females when they're sitting on eggs. Chicks have a pale brown down with stripes, providing them with excellent camouflage as they forage across the forest floor.

Pigment

Tiny granules of coloured compounds are utilised by the bird's body to give the feathers their unique colour characteristics, and depending on how they're aligned within the feather light can be refracted in different directions to give different colours; and how light is reflected can make the colour appear different again. There are several genetics-based books on the market that are ideal if you have an interest in this field.

The only two basic colours present are red and black, with all the other colours that you see being variations on this theme, due to the interaction of genetics. Patterns are produced by altering the growth rate of the feather while a certain colour is laid down, for example a check in growth as a result of either an external variable such as temperature, or an internal genetic factor, can cause an otherwise pencilled feather to become barred or an otherwise black feather with a green sheen to develop a purple sheen. In addition there are pigment colour-changing genes, for example the 'silver' gene that changes the colour red to appear as silver.

Once the chicken's genes have decided how black and red will develop, another set of genes come into play that decides where and how the colours will be displayed, and may well alter the resulting colour again – so we see orange/ red, bay, buff, tan, mahogany, browns, yellows and salmons coming into existence, along with colour restrictions to saddle, breast, tails hackles or wings, to name a few.

Patterns

There's so much variety to be found in the patterns of chicken feathers that not all of them can be listed here, and even within breeds there are minor variations according to tradition and breed standard. Consequently what follows is intended as a guide to those that you're most likely to come across rather than a definitive list:

Barring
Barred feathers have distinct transverse (horizontal) bars of two different colours across the feather, for example in the

Above: cuckoo.

Barred Plymouth Rock. This is caused by an interruption process that stops colour being laid down as the feather grows. The pure male with two barring genes produces twice the amount of pigment inhibitor, therefore the feather is lighter, as it will have twice as much white as that of the female. Barring is sex-linked, with male chicks exhibiting a head spot (thus making it easy to separate the sexes on hatching), and is a useful trait both when investigating genetics and for the breeder who's only interested in chicks of a specific sex. Irregular barring is often referred to as cuckoo, where the tips of the feathers are also dark.

Mottling
This is similar to barring in that the colour is inhibited. The difference, however, is that only the tip of the feather is affected, producing a white end as, for example, in the Ancona. Other patterns caused by the mottling gene are speckling, mille fleur, and apical spangling. Only one in two to five feathers is affected.

Spangling
This produces a tear-shaped spot on the tip of an otherwise solid feather, as in the Silver Spangled Hamburg. Females exhibit a larger spangle due to the action of oestrogen on the formation of the spot.

Lacing

There are generally two types of lacing, single and double. In the single variety there's an outer ring on the edge of the feather (border) and it appears in both sexes, although in the male the pattern is typically restricted to the breast and ventral feathers. Birds that exhibit 'henny' type feathering, *ie* feathering typical of a female bird (for example the Sebright), exhibit the pattern all over the body.

Double lacing is simply two rings, one inside the border ring with a space of background colour between them. This is typically only exhibited in females, with the males usually of 'wild type' colouring.

Pencilling

Pencilled feathers have bars across them, and can be horizontally or vertically striped. They're typically a female feather pattern and should have three or more pencillings.

Other colours

Besides the normal colours associated with chickens, there are a whole host of combinations that can come together to form colours you may not have heard of:

■ Mahogany – a deep red/brown.

Above: Buff lacing (Chamois).

■ Porcelain – describes straw-coloured feathers with white tips and a pale blue stripe through part of the feather.
■ Birchen – black with silver-white feathers, almost a Columbian in reverse except that the tail is the basic colour of the body. The hackles of both the male and female have silver-white feathers, as do the saddle, back and shoulder coverts of the male. The male neck hackle

Left: Mahogany is a rich red.

Left: Crele may be referred to as tolbunt.

has a narrow black stripe through the middle of an otherwise silver-white feather. The breast feathers are laced silver-white in both sexes.

- Partridge – the colour that you associate with a typical cock: its tail and breast are black, with orange-red hackles and saddle. The wing bows and back are a brilliant red with a glossy green bar across the wing, with brown wing tips. The female expression of the colour is much drabber, with the majority of the feathers a bay/brown colour with a deep salmon breast. The bay feathers may be marked slightly either with lacing or peppering (fine black spots). The black markings on the wing bows may be altered by genetic factors, giving rise to blue partridge, golden partridge etc.
- Salmon – a rich pinkish buff.
- Crele (or creole) – a chequered (barred) orange.
- Duckwing – the head and saddle of the male are coloured, which gives the 'duckwing' its colour; for example, a silver-white colour gives silver duckwing, and yellow gives a golden duckwing. The wing bow and shoulders also take this colour, which is where the name comes from. The breast and thighs are black.
- Bay – a soft pale brown.
- Columbian – also referred to as 'ermine', this has black restricted to the tail and hackles, typical of the Light Sussex.
- Pyle – a complicated series of colours where, in the male, the hackle is a bright yellow through to red, with the back, saddle and wing bow a richer shade of the same colours. The wing secondary feathers are deeply coloured and this

should show when the wings are closed. The female has a white hackle, tinged with colour, and a rich salmon breast. The remainder of both sexes is white.

- Quail – similar to partridge and typical of Belgian Bantams. The quail-coloured male is a more golden colour with black hackles and tail. The saddle is black with golden brown lacing, and the wings and shoulders are golden brown. The breast of a quail-coloured female is a warm gold-brown.
- Wheaten – a golden orange or yellow, depending on breed.
- Polecat – an otherwise black bird with dark tan streaks on the hackles and saddle in the males.

Below: Columbian restricts black feathers.

Breeding

Reproduction

When the hen or hens have completed a clutch of around 11–13 eggs, one of the females will commence sitting tight on the nest and will become 'broody', a term used to describe the sitting process. An odd number of eggs tend to fit together better than an even number, making them more secure in the nest. A broody hen becomes fiercely protective, not only to ward off predators but also to stop other hens from attempting to lay in the nest and thereby cause incubation problems.

Development of the egg only begins when the hen sits – an aspect utilised by poultry enthusiasts for shipping hatching eggs – and results in them all hatching within a day or two of each other even though it may have taken many days to lay the clutch.

The hen turns the eggs regularly during incubation to help the exchange of gases and to keep the embryo mobile, and she very rarely leaves the nest even to eat and drink in order that a stable temperature and humidity level can be maintained. However, she can leave the nest for up to 20 minutes without any harm befalling the eggs.

After approximately 18 days the chicks begin 'pipping' (breaking through) into the air space at the blunt end of the egg with a protuberance called the 'egg tooth', situated on the end of the beak, and the hen begins to cluck gently, stimulating the chicks further.

Having broken a small hole through the shell, by day 21 the chicks rest for a while to absorb the remaining egg yolk and withdraw the blood supply from the shell membrane before working around the shell to enlarge the hole and pushing the blunt end of the shell off with its feet.

Wet and shaky, the newly hatched chick soon dries off and finds its feet, safe under the protection of the hen, who'll usually stay on the nest for a couple of days more after hatching to ensure that as many chicks have emerged as possible.

Above: Assemble your breeding pens with birds possessing qualities that you wish to preserve.

During this time, the newly hatched chicks will live off the absorbed yolk. Any eggs not fertilised by a male or made infertile by genetic factors won't hatch, and the hen eventually loses interest in these and leaves the nest.

The new mother will jealously guard her young, and although she may be helped by the male the task of brooding the chicks falls to her, and she will hover over them using her skin and feathers to keep them warm and safe. The chicks don't stray very far for the first couple of days, venturing out only when called by the hen when she sees an edible morsel – she issues a quick 'chuck, chuck, chuck' call and often picks the morsel up and drops it in front of the inquisitive chicks. When the young have developed their harder feathers after a couple of weeks, hormones in the hen's body signal a return to laying and she loses interest in the brood, who are left to fend for themselves but don't stray too far from the flock or their mother.

Modern egg-laying breeds rarely go broody, and those that do often stop partway through the incubation process, making them unreliable. Certain pure breeds, notably the Cochin, Cornish and Silkie, do regularly go broody, and make excellent mothers, not only for chicken eggs but also for those of other species such as quail, pheasants, turkeys and geese.

Mating

There's one main drawback to consider when mating your birds, which is that half of their progeny are likely to be male, which means that you need to be prepared to dispose of 50 per cent of the youngsters. If you can't handle this then it's not advisable for you to breed your own birds, as this responsibility comes as part of the package.

January is one of the best times to start mating, since by the time the chicks have hatched and become independent of a heat lamp (see page 115) it will be nearly March, when the weather is just warming up and the youngsters can be put out on to grass.

The first thing you will need to do is assemble your breeding flock (also termed a breeding pen), usually consisting of one cock and several hens, as the cock will mate or 'tread' several times a day and if a single hen is run with him her life can become a misery from his over-zealous attentions – you may find feathers scraped off or, especially with the larger breeds, wounds appearing, which are best sprayed with gentian violet spray. In such cases a poultry saddle should be fitted. Some birds, however, such as the game breeds, can be kept in pairs and often remain faithful to each other.

The maximum number of birds running together will depend on the size of the breed, with the more placid ones, such as Orpingtons or Silkies, having two to five females per male, while more active breeds such as Leghorns have more.

Below: You can keep a number of males together if there are sufficient females, but keep an eye out for any serious fighting.

If you're in any doubt regarding numbers you'll soon see if all the girls are being serviced or if your male has a favourite. If you find fertility is low, then it may be worth considering fewer hens, although a couple of successful matings per week should be all that's required, as sperm can remain active for over a week.

After putting together the pen, leave the birds to get used to each other. Often a hen will not accept a new male, however ardent he may be, so leave them alone and undisturbed for as long as you can, ensuring that they're well fed and comfortable. Generally you can be sure of viable eggs after about two weeks, providing the male is fertile. Infertility in cocks is often temporary and may be linked to light availability and temperature; if the male is too fat or undernourished his fertility will also suffer. Older birds may be fertile in the spring and not in the autumn.

Before mating a rooster may call the hens over to him by clucking in a high pitch and picking up and dropping small items found on the ground, imitating feeding behaviour. In some cases the male will drag the wing nearest to the hen on the ground, tilting sideways while circling her, which is all part of the courting ritual. When a hen is used to coming to his 'call' the rooster may mount her and proceed with the fertilisation.

The egg

An egg is the perfect package for reproduction, consisting of seven basic parts: the shell, membranes, the white (albumen), the yolk, the chalazae, the germinal disc and an air sac.

The shell is composed mainly of calcium carbonate taken directly from the hen's body, and is an ovoid shape allowing it to roll in a circular direction, meaning it will never roll far from the nest. The shell is porous to allow carbon dioxide and moisture to escape in a regulated fashion from the developing chick while at the same time allowing the free ingress of oxygen needed by the youngster. The external surface of the shell is protected by a 'bloom' deposited on the egg as it's laid, which provides a natural barrier against the ingress of pathogens – a reason why washing eggs is discouraged, as it removes this layer. Being naturally thinner on the inside and used by the chick in its development, the egg is easier to break out of than into, a particularly useful trait in the wild.

The two membranes form the air sac through which gases diffuse and which grows over time to provide an air space through which the chick's beak passes on 'pipping'

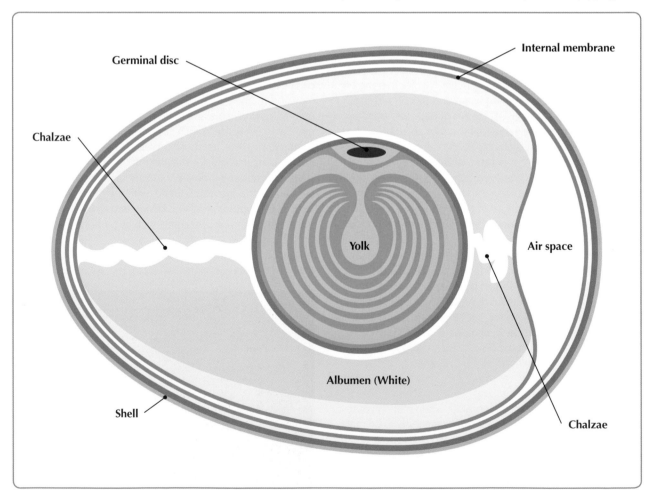

Germinal disc

Internal membrane

Chalzae

Yolk

Air space

Albumen (White)

Shell

Chalzae

Formation of the egg

Each hen is born with a full complement of eggs in her body, and depending on breed will lay a few over many years or a lot over a few years, and once they're used up she'll lay no more. The ovaries are situated at the back of the abdomen and release a fully formed ovum with a yolk sac attached into the neck of the infundibulum, which is a funnel-like opening to the top of the uterine passage. Just below the infundibulum is the oviduct, where fertilisation of the ovum takes place when it's fused with the sperm from the cock-bird, which in some breeds can stay in the oviduct for up to three weeks after mating.

Cell division begins immediately, and as the fertilised egg travels through to the magnum it is covered with albumen, and then the membranes in the isthmus. The shell is added in stages in the uterus and finally egg shell markings and colour are added before it moves through the vagina and is expelled by the cloaca, a separate organ used to push the egg out past the vent opening and prevent contamination by faeces.

Incubation

Above: Coiled chalzae are easily visible.

to enable it to breathe inside the otherwise fluid-filled egg. One membrane is secured to the shell with the second loosely secured to it, allowing it to separate at the broad (blunt) end to form the air space as the egg loses moisture through evaporation. As the egg ages, regardless of fertility, the air space expands, which is a property utilised to check for the age of an egg, as it will float blunt end up if old and past its best.

The white is made up of vitamins, minerals, protein and water and consists of three parts: outer and inner watery parts, and a thicker middle part that provides cushioning for the embryo as the egg is moved around. The outer layer facilitates the movement of gases in and out of the egg, while the inner layer allows the embryo free movement, enabling it to come into contact with food and oxygen.

The yolk, typically yellow, is made up of vitamins, minerals, protein, water and fats, and is used as a food store for the chick on hatching as opposed to while developing, being absorbed into the body wall a couple of days prior to emerging.

On top of the yolk is the germinal disc, also called the egg cell, which is the ultimate powerhouse of the egg. It is here that the embryo will form by a process of cell division and growth once fertilisation has taken place.

Chalazae are strings of twisted tissue, similar in appearance to thick cotton, whose purpose is to stabilise the yolk and germinal disc when the egg is moved, and to prevent twisting and misalignment. It's important to remember, when turning eggs, not to turn them constantly in the same direction, as this would result in one chalaza coiling tighter while the other uncoils, destabilising the embryo.

Incubation can be either artificial or natural, with both processes having their champions.

Natural incubation occurs when the hen sits, and poultry keepers have for years made use of broody hens to carry out the process when they needed it rather than when the hen decided to.

If you want to incubate your eggs the natural way, you'll need to have selected a hen, preferably a Silkie cross, and set up a house in which she can sit undisturbed. A Silkie cross is a much more reliable animal than a Silkie, since the latter will sometimes sit until she dies of malnourishment or thirst, such is her propensity for brooding. In addition, a Silkie's dense under-fluff has been known to be lethal if a chick gets tangled in it. So it's much better to have a cross, which has much of the same mothering instinct along with harder feathers and a greater sense of self-preservation.

Traditionally a turf was cut from the lawn and placed soil side upwards in the nest to provide humidity, and older breeders still swear by this practice.

Ensure that the hen is serious about brooding, as often they'll sit for a couple of days and then give up. You can check by slipping some crock eggs under her, and if she cuddles your hand clucking and pecking it's likely that she's settled. Once you're certain of her intentions you can introduce your collected eggs. Throughout the brooding process you should get the hen off the nest periodically to defecate, eat and drink. If she messes near the nest you'll need to remove it.

Artificial incubation was practised for centuries by both the Egyptians and the Chinese, who handed the knowledge down through the generations. They made use of fires to maintain the correct temperature, the Egyptians alledgedly placing the egg against an eyelid to check that the temperature was right.

Europe didn't really catch on to artificial incubation until the 18th century, when poultry-keeping became popular as a hobby rather than a way of life. The earliest method utilised rotting manure to provide a heat source. Fortunately, with the Industrial Revolution well under way it was not long until the first fully artificial incubators began to appear, one of the first being shown at the Great Exhibition in 1851. With the development of thermostatic control artificial incubators began to take on a more familiar appearance, and with paraffin as a readily available power source the stage was set for incubation as we know it today.

Modern incubators are technologically as good as, if not better than, a broody hen, with fewer variables that could lead to a poor hatch. A top of the range model will give you little change from £1,500, but will hatch almost anything you care to put in it, and requires little more expertise than knowing where to fill up the water reservoir. It could be argued that this has removed natural selection, as otherwise inferior chicks may be hatched which under natural incubation would have died. However, the other side of the debate is that extinct or endangered breeds may be preserved for future heritage. Whatever your reasons may be for using an artificial method, the same basic rules apply.

Choosing an incubator

Your choice will depend to a great extent on how many chicks you wish to hatch, the reason for hatching, and the amount of money that you have at your disposal. Consider, too, the power source that you have available, and where you wish to put the incubator.

Above: Smaller incubators (here with an automatic turning cradle) will hold 10-12 eggs.

Before committing yourself to a particular type, consider your situation and what you want to get out of hatching. The correct decision now will help towards your incubation success in the future. If you only want to hatch a few eggs then a tabletop incubator is probably the best option, especially if you want children to be able to see into the apparatus as the chicks hatch. These are a very manageable and relatively cheap option, and ideal if you're just starting out.

Your next consideration is likely to be whether you want to take control of turning the eggs or leave it to the machine. With manual turning you must be willing to perform the operation regularly, otherwise your hatch rate will suffer.

Below: Larger table top incubators will hold around 40 eggs.

It's much better to leave it to automatic turners, although you should check regularly that they're working.

Air circulation is necessary for good gas exchange and to pass heated air over the eggs. For this there are two types of incubator, still and forced. Still air incubators rely on warm air rising and cooler air sinking (convection) to generate a flow over the eggs. These incubators are typically cheaper but lack a lot of functionality unless you only intend to hatch a few eggs. Forced air (also termed fan-assisted) incubators drive the warmed air over the eggs and are a more adaptable, if more expensive, alternative.

Automatic controls consist of a thermostat that must always be built in to regulate the temperature within the incubator, turning the heat off when the selected temperature is achieved and turning it back on again when it drops. In the more expensive models, you can elect to have electronic controls for temperature and humidity, which make things a whole lot easier still. Be warned, however, that reliance on electrical functions makes it absolutely essential to have the incubator regularly calibrated; and if things go wrong, you're unlikely to recover even a small percentage of your hatch.

Separate areas for hatching and incubation is a useful option if you want to hatch in a constant stream, especially for exhibition or commercial purposes as the eggs are moved to a separate area in the incubator where turning stops and the eggs are allowed to hatch. This enables the damp

chicks to wander around while they dry and build up their muscles as they explore the new surroundings. In all but the largest incubators, it is difficult to provide the different humidity and temperature requirements of the hatching eggs, which is why those incubators designed to be both hatcher and setter should be operated to the manufacturer's specifications.

Be sure to read all the instructions that come with the incubator, and if in any doubt whatsoever consult the manufacturer or dealer.

Incubation requirements checklist

- Incubator, serviced and tested at least one month before you need it in order to allow for any spare parts to be ordered and fitted.
- Scales for weighing the eggs (to gauge weight loss).
- Candling lamp or box for internal assessment of the eggs and developing chicks.
- Antibacterial spray for cleaning equipment and antibacterial gel for cleaning hands.
- Spare thermometers, one for the room, one for the incubator.
- Hygrometers (to measure humidity), one for the room one for the incubator.
- Record book.
- Pencils.
- Rubber gloves.
- Dehumidifier (if needed).
- Reference charts (eg for embryo development).
- Backup generator if electricity supply is unreliable.
- Extension cables (if you have insufficient or badly placed sockets).
- Incubator instructions.

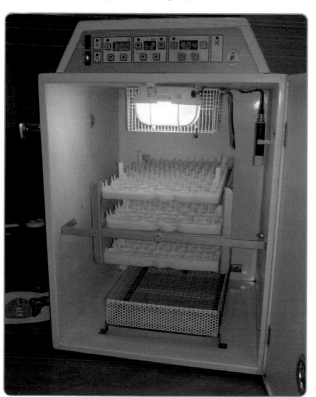

Below: An automatic incubator is the most accurate if you can afford it (note the seperate hatching tray at the bottom).

Below: Incubator instructions must be followed for successful hatching.

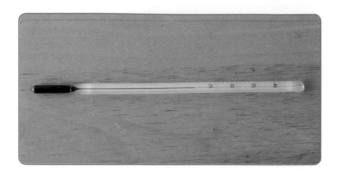

Above: A thermometer is an important piece of incubation equipment.

The process

In order to maximise your hatching success you'll need to stick to a strict procedure, and over time you can tweak what you do to improve things further. Remember that hatching is an art – after all, it took the chicken thousands of years to perfect it, so don't expect to be an instant expert. The key requirements are as follows:

- Make sure the eggs are clean, but wash them only if necessary (see below).
- Store in a temperature of 15-18°C, and set within seven days.
- Incubate at 37.5°C (or as per the incubator instructions).
- Maintain an external humidity of 30-60%.
- In the hatcher, maintain a humidity of 75%.
- Ensure adequate ventilation.
- Turn regularly (five times a day).

Below: A syringe will help you fill a small reservoir.

Because you're dealing with an artificial environment, the natural skills of the hen aren't available to you, and you have to adapt the process accordingly. Keep all equipment thoroughly clean and disinfected and wash your hands before and after handling the eggs. Clean dirty eggs before storing and setting.

Adequate ventilation is important to ensure good gas exchange and to maintain a satisfactory level of humidity. Most incubators will have ventilation holes that can be opened or closed as appropriate, and this will be explained in the instructions. As the chicks grow, so does their need for oxygen, and this is achieved by opening more of the ventilation holes. Poor ventilation will mean a build-up of noxious gases that will damage or kill the developing embryo.

The temperature needs to be kept stable, something that a hen will achieve along with humidity control by the contact of her skin with the egg and regular moving. The ideal temperature at the core of the egg in the incubator will need to be 37.5°C (99.5°F). This should be measured with a calibrated thermometer. Remember to check the manufacturer's specific instructions, as still air incubators may have colder areas, and the temperature measured above the egg may need to be higher to ensure the correct core temperature. With fan-assisted incubators it's likely that the whole of the inside can be maintained at 37.5°C, but again, be sure to check the instructions.

Run any equipment for 24–48 hours before setting to ensure that no problems are evident, and in any case ensure that your incubator is serviced and tested well in advance of beginning the season. Slight changes in temperature may severely affect the embryos in the first

couple of days, so a maximum/minimum thermometer is excellent for checking that the temperature has remained stable while you're asleep.

Next to temperature, humidity is probably the most important single factor as it helps to regulate the amount and speed of gaseous exchange, keeps the egg moist, and provides the air space that the emerging chick first breaks into. Too much humidity means that oxygen exchange is reduced and the chick will die, too little and the membranes dry so that the chick has trouble breaking out and dies. Most incubators have a water reservoir that will need filling at some point in the process, and if the relative humidity in the room is at 50% then you may need to limit the amount of available water at a specific time in the process; only experience and a knowledge of your specific conditions will allow you to achieve maximum success for your particular conditions.

Egg collecting

This should be done at least twice every day to ensure that eggs are fresh, unsoiled and not sat on, as the hen's body warmth will start development of the embryo which is then stopped by you removing the egg and storing it, which will result in the death of the embryo (indicated by a 'blood ring' when the egg is candled). Use a basket or soft container to carry the collected eggs, and as soon as you pick them up you must mark them with the date and any other details you may need, such as breed or pen number.

Your hands should be clean before you handle the eggs, as you may otherwise unwittingly pass pathogens on to them. The gel cleansers sold in most pharmacies are ideal for this – the gel evaporates from your hands leaving them dry without the need for a towel, and you can carry a small bottle around with you.

Below: Eggs should be identified before incubation, and a record kept of what the ID means.

Above: Clean eggs carefully and only if necessary.

Cleaning

Unless the eggs are excessively soiled or you're incubating a rare breed that you can't leave to chance, *don't* wash them, as this removes the protective bloom that prevents pathogens getting into them. Surface dirt can be wiped or picked off. However, if you must wash them then do so with a specialist egg sanitiser in a lukewarm wash of water, and follow the directions for making up the solution carefully.

Warm water will cause the air inside the egg to expand and will push any microbes outwards rather than pulling them inwards, which is what will happen if you use cold water. It's useful to put a thick cloth in the bottom of the container that you're washing them in so as to reduce the risk of breaking eggs if they're inadvertently dropped; and change the water regularly if you're washing a lot of eggs, to stop it from going cold. Ideally you should place the damp eggs in a specially designed carrier (readily available from incubation specialists), at least until they're dry.

Storage

Because you want to slow the process of cell division as long as possible, store the eggs in a cool atmosphere at around 12°C (53.6°F) and turn them end over end at least three times daily (left to right then right to left on the next turn and so on). This will prevent noxious gases building up within the egg and allow a constant supply of fresh oxygen to the germinal disc, while also keeping the embryo mobile.

Fertility decreases after day one, so you should start incubation before seven days to ensure the best hatch rate. To prevent a shock to the embryo allow the eggs to come up to room temperature before setting them in the incubator.

Quality

The eggs that you select should be undamaged, free of hairline cracks (use a candling lamp to check) and not deformed in any way. However, if the breed you're

Above: Allow eggs to dry after cleaning.

incubating is very rare you may have to use the eggs anyway – specialist incubation companies can supply a special egg glue to cover the cracks.

The shape of the egg is an important factor as a deformed egg may have a displaced air sac, and the emerging chick will have difficulty on hatching and may drown if unable to pip into this space.

Shell texture is important too, as a chalky shell may be porous and let in microbes, while a shell with a thick band around it – possibly caused by a shock to the hen – may make it more difficult for the chick to emerge.

Setting

This is the process whereby the eggs are put in the incubator to start their development. While many large hatcheries have a separate area for this, which is maintained at around 24–28°C (75.2–82.4°F) and a relative humidity of

Below: Eggs should be loaded on their side if you cannot place them blunt end uppermost.

60%, the average poultry keeper doesn't have the luxury of separate setting facilities, and often has to use the spare room to store, set and hatch his eggs. By keeping the temperature in this room as stable as possible you lessen the fluctuations in temperature that the incubator will suffer and therefore improve your hatch. You can do this by ensuring that the room is draught-free and keeping any radiators on a constant low. If strong sunshine is a problem then draw the curtains.

Site any incubating equipment in a part of the room where it can't be knocked, as this too will affect the incubation process.

Loading

With the incubator warming up and any humidity controller switched on, you should bring the eggs in from the storage area to allow them to come up to the temperature of the room. Ensure that they're marked up with any information that you'll need to record, for example breed and date collected. It's useful to mark the shell with a cross if you have no other data to log, so that you can check that the eggs are turning correctly in the incubator.

Once the incubator has come up to temperature, introduce the eggs, either by laying them on their sides if you have a small machine, or blunt end upwards if you have larger trays to load them into. The chick will emerge into the air space that forms at the blunt end, so placing them the wrong way up (ie pointed end upwards) may kill the chick. Smaller incubators have rollers which roll the eggs from side to side, while larger models tip the eggs at 45° angles from left to right.

Don't mix different species, as their hatching times and temperatures will differ, and remember that true bantam breeds may emerge a day or so ahead of larger breeds, even though the conditions will be the same.

Take care as you load the eggs, especially in trays, as you may be working at an awkward angle and a sudden jolt can crack the shell or displace the embryo, causing developmental problems. If you're using an automatic turning device, switch it on.

Candling

Your routine for the next 21 days should be fairly straightforward, with regular candling to ensure that the chick is developing normally and that the air space is growing to plan. Candling takes its name from the practice of holding a developing egg to a candle flame to check its progress. Nowadays there are much better options available, from hand-held torches to light boxes, all designed to illuminate the contents of the egg so that you can check progress. An egg needs to lose between 11% and 13% of its initial weight in order for the air space to be correctly developed. Too little loss (high humidity) and the

Above: A candling lamp is a necessity.

air space won't be big enough, and the chick will drown. Too much loss (low humidity) and the air space will be too large and the membranes will dry. High humidity is more of a problem than low, and you may find the use of a dehumidifier in the room cures the problem.

At around seven days you'll be able to see if the embryo is developing correctly, and any eggs whose development is abnormal should be removed to prevent them going off. A non-fertile (clear) egg will begin to decompose and build up noxious gases until the pressure causes it to burst (termed a 'banger'), contaminating the good eggs as a result and introducing disease.

You should be able to easily identify a fertile egg as it will have a central blob with veins radiating outwards when candled, similar in appearance to a spider with long legs. If the egg is clear apart from a red ring around the shell, it's likely that it was fertile but has died, either as a result of lethal genes (where a particular genetic combination results in death) or for some other reason, most often partial incubation before collection (if, for example, it was sat on by a broody). This is why it's important to collect eggs at least twice a day, as mentioned earlier, or even more frequently if you know that there are broody birds around.

Darker eggs such as those of a Marans may need candling at 14 days to ensure fertility, as their darkness makes them fairly difficult to see into.

Hatching

At around 18 days, you should turn off any turning equipment and allow the eggs to settle, as the chick will be in the process of manoeuvring itself round to break through into the air space and will become disorientated if the egg continues to turn. This breaking-through process is known as 'pipping', and involves piercing the membrane and taking the first gulp of air before breaking through the shell with a specially adapted 'egg tooth', a pointed protuberance on its beak which falls off shortly after

Below: Only help the chick if no progress is made after a couple of hours.

Left: Unhatched eggs (here on the left) can be left in the incubator for a further 48 hours.

hatching. At this point you should raise the humidity and lower the temperature, which has the effect of increasing humidity still further in order to keep the membranes wet and aid the hatching process.

If you have a separate hatching area, be it in another machine or the bottom of the incubator, the eggs will need moving gently into their new location. If you're using a divider to separate the hatching offspring from different parents then this should be put in place now. Avoid, if you can, putting the eggs in little bags to segregate the chicks, as these are not practical, even though some breeders will swear by them.

An emerging chick will begin to chip round the middle of the shell and, using its body and legs, gradually force the two halves apart until it is free and is able to take a rest. At this stage the wet chick will be utterly exhausted and will lay quite flat and still until it has recovered. During this time other chicks will also emerge, and things can get very hectic in the hatching tray, but don't be tempted to interfere at this stage as the chicks are susceptible to temperature changes.

As regards helping out a chick in difficulty, breeders are divided in opinion. Some say that it's best not to help as the chick will likely not be the healthiest of specimens, while some argue that artificial incubation is just that, artificial, and any help you can give is OK. If you must help the chick – for instance if no progress has been made after a couple of hours – check to see if the membranes are dry, and if this is the case wrap the egg in a cloth moistened with warm water containing egg sanitiser to rehydrate them. Then gently pick a little of the shell away from the hole in a direction that follows around the middle of the shell. If the membranes bleed, replace the egg in the

incubator as it isn't ready. Repeating this every few hours should enable the chick to absorb the blood from the membranes and will enlarge the crack until it can emerge safely.

The chicks should be left to dry in the incubator until they're fluffed up and toddling about the hatching tray, where they can stay up to 48 hours as the yolk absorbed into their bodies will sustain them. Remember, however, that any deformed chicks should be humanely destroyed with a hand-held pair of dispatching pliers.

Remove the chicks to the brooder and clean out the incubator thoroughly to remove all shell remnants, droppings and down fluff.

Below: The newly hatched chicks can be moved to the brooder unit 48 hours after hatching.

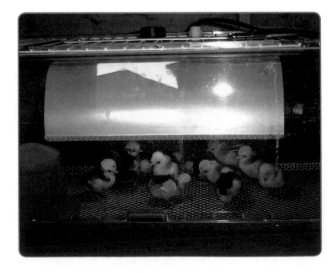

Brooding and rearing

Brooding is the process whereby the newly hatched chicks are grown on in a protective and warm atmosphere until they've developed their harder feathers at around six weeks. Normally this would be achieved by the hen huddling over her chicks, keeping them warm and safe. However, in an artificial environment you'll need to provide a heat source, food, water and protection, and as with the incubator, you'll need to get this all ready in plenty of time.

The brooder area is typically a box with a heat source suspended over it and a layer of shavings on the floor. Some breeders will use a tea towel in the bottom for the first couple of days, to allow the chicks to get a good purchase on the floor in order to strengthen ligaments and tendons so as to prevent splayed legs.

Mount a chick drinker and feeder on small blocks just high enough to allow the chicks to get access while at the same time preventing them from soiling the reservoir. Add a probiotic to the drinking water, and as you move the chicks in dip each one's beak into the reservoir to ensure it gets a drink. This ensures that good gut bacteria can flourish and the chick will get an excellent start. Purchase

a good quality chick crumb with a protein content of 17%. Finally, ensure that the area is safe from rats, as they will kill chicks in their hunt for an easy meal.

The heat source is typically a heat radiant lamp, either ceramic or infrared (light-emitting lamps should be avoided, as these cause confusion and may lead to early bullying behaviour such as feather-pecking, and the chicks will need to rest at some point, which they will find hard if constantly active). The lamp is suspended over the brooder at a height which gives a temperature of about 34°C (93.2°F) initially when a thermometer is placed about 100mm (4in) above the floor.

The chicks will give you a good indication if the lamp is positioned at the wrong level, as they will huddle and cheep if it's too high (temperature is too low), or sit beyond its reach, sometimes panting, if it's too low (temperature is too high). You should aim for a drop of about 2°C per week (more if it's warmer outside) until the lamp can be removed. This is typically at six weeks of age (although it may be longer, depending on time of year and breed), by which time the chicks are fully feathered and old enough to survive without extra warmth.

Once the chicks are independent of the heat source you can consider moving them outside into a rearing house, although you'll need to make sure again that no rats can gain access. The house will need a solid run attached to prevent escapees and to provide security for the growing brood; some breeders attach a wire floor to achieve this.

Below: Chicks will head for the warmth under the lamp.

Above: An infra red lamp is the best option.

After six weeks you can move the chicks on to a growers' ration by gradually mixing a handful into the chick crumb, and steadily increasing the amount over a week so that after seven days the chicks are eating growers' pellets only. The reason for this is that chickens are naturally wary of changes to their environment and a sudden change in the appearance of the food can cause confusion and stop the chicks eating, something which at this stage you don't want to do, as they can then become prone to disease through stress.

Provide an ample supply of fresh water and allow access to grass and the chicks will begin to put on size and weight as they start to forage.

The leg-ringing scheme

The Poultry Club of Great Britain's leg-ringing scheme began in 1994 with a lacklustre take-up, but support has since increased considerably and 30,000 rings were sold in 2008. The rings, which are purchased in multiples of ten, help in the identification and recording of breeding lines, making them a crucial tool in the conservation and preservation of all the pure breeds of poultry. Each is individually numbered with the breeder number and the bird number plus the current year, and each year a different colour is used.

As security is an important aspect of breeding and showing, ringed birds can be traced as long as the ring is intact, and even when a bird is sold the ownership can be transferred through the Poultry Club's record system. In addition it was recently agreed that the ring number may be tattooed under a bird's wing as an extra security precaution.

The ringing process is simple. Once the hatchlings have been identified – normally by using small wire cages with divisions in the hatcher, to allow the eggs to be sorted at 18–19 days – the chicks are subsequently ringed with expanding chick leg rings. They are then ringed with the permanent ones at around ten weeks, according to breed. Once ringed, information is written up into a book and a permanent record achieved. As culls are made, the rings can be recycled for later hatched chicks and the records adjusted accordingly.

There is initially a degree of trial and error, and timing is crucial with some stock – for example Sussex bantams, where the cockerels need a D size ring and the pullets a C. Pullets need to be ringed two to three weeks before the cockerels and the odd one may get missed.

If you would like to participate in the Poultry Club's ringing scheme you can contact them by email at

Below: Numbered leg rings are issued by the Poultry Club of GB.

info@poultryclub.org, where you can discuss your requirements with the Ringing Scheme Co-ordinator.

If you're going to use leg rings, you should order them in October for the following season's hatch.

Vaccination

Vaccines come in two types, live and deactivated, and can be administered in the drinking water or by injection using an insulin syringe or a calibrated gun.

Certain breeds such as Silkies are prone to Marek's disease, a herpes-type virus that infects the brain causing, amongst other things, paralysis and tumours, and breeders will regularly vaccinate chicks at a day old and again at two weeks to improve immunity. Vaccines are also available for other diseases including Newcastle disease (ND) and infectious bronchitis (IB), and should be administered at the correct time during the chicks' growth – ND is typically administered at two weeks old and IB at four weeks old.

However, there's a school of thought which would argue that vaccination weakens the natural immunity present in a line of birds, and some breeders will attempt to breed immunity into their stock. The current recommendation is that unless you have a problem either on your property, in the breed, or in the immediate area, it's best *not* to vaccinate.

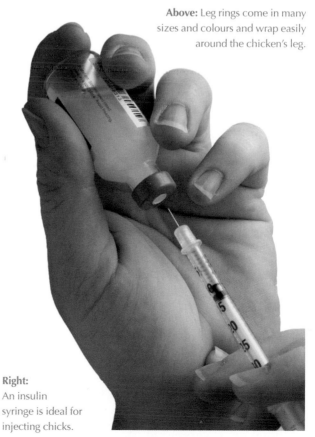

Above: Leg rings come in many sizes and colours and wrap easily around the chicken's leg.

Right:
An insulin syringe is ideal for injecting chicks.

117

Showing

Above: A few good birds can soon fill a trophy cabinet.

Above: Birds are judged to pre defined standards.

Showing chickens is a popular hobby that can be enjoyed by all the family – all shows have prizes for junior keepers and many have a 'best ladies' exhibit' section. Although some of the local shows will have a general section for children where you'll be able to enter a hybrid, all exhibits will otherwise be pure breeds of a type recognised by the Poultry Society of Great Britain, who, with the help of the individual breed clubs, will have drawn up a standard of perfection for each breed. It's the points allocated for each feature that decide the quality of the bird.

It should be remembered that there are many ancient and valuable lines of chickens, just as much as there are for dogs, cats and cattle, and these lines can trace their history back many hundreds of years. So, needless to say, there's a lot of pride and competitive spirit within the show fraternity, and if you want to compete at National level you've got a lot of hard but extremely enjoyable work ahead of you.

Each breed will have a breed club, which you should join to get the latest help and information regarding your chosen breed. It will familiarise you with the standards applicable and get you introduced to the secretary, chairman and committee, who'll be able to advise you as you get started.

Shows are run to strict guidelines and judges are chosen a year in advance and will either be qualified to Poultry Club standards or be experts in their chosen breed. However, it's worth knowing your judges too, as even though they'll deny it they do have preferences, and a bird not placed by one judge will be on Championship Row when judged by another.

History

Showing originally started as a way for poultry fanciers to compete following the abolition of cockfighting. The best and fittest birds were compared with each other and were awarded points for their 'standard', and how closely they conformed to the characteristics of their individual breed.

The first official standards were introduced in 1865 to maintain uniformity for the few breeds being exhibited. Then at the turn of the century enthusiasm for showing brought in many breeds from the Continent and America, and exhibition as we know it began, with the Poultry Club becoming the guardian of the standards. Shows sprang into being up and down the country, with huge prizes for the best examples of breeds, and stock was shipped up and down the United Kingdom via the rail network to compete for the prestigious prizes that would bring fame and a good income to the winners.

Some of the original trophies are now stored in secure locations and winners may never actually get their hands on the real thing, being issued with a photograph instead.

Preparation

Anyone interested in showing should do their homework well in advance by going along to a local show to see the standard of birds on display and talk to the exhibitors and judges, who'll be more than willing to give advice if you make it clear that you want to learn, and will at least have

Above: You can often spot a potential winner from an early age (here a young Japanese bantam).

a basic understanding of the questions that you'll be asking. If you're really lucky and demonstrate your enthusiasm an exhibitor may even take you under their wing and teach you all that you'll need to know. Read as much as you can and invest in a copy of the *British Poultry Standards* for reference.

One immediately noticeable difference to showing dogs or cats is that normally the stock you exhibit has been bred by you rather than bought in, although this does occur. It's always best to obtain parent stock that conforms to the breed standard, preferably from an established breeder who you'll have met during your initial investigations. The breeder will have a better idea of the birds that will produce good

Below: As soon as the sexes can be distinguished it is best to split growers into separate pens.

offspring, and you should be guided by their experience and advice, as it's not always the best-looking birds that produce show winners.

Serious show preparation starts a year before the exhibition date with the breeding stock and growing on of the chicks, which is hugely exciting as you watch them develop and begin to see the best specimens standing out from the others. By knowing your breed, you'll have a good idea of how long they take to mature and therefore the amount of time that you need in order for your birds to reach peak fitness and appearance at show time.

Males will stay in condition longer than females, who'll become drained by the rigours of laying. Some exhibitors will even keep their pullets and hens on whole wheat only, effectively reducing their nutrient content and thus delaying laying until after a show. Laying can also be delayed by moving the birds from pen to pen, unsettling them and halting their desire to nest.

However, feeding your birds correctly is essential. They should be fed with a good-quality chick crumb, growers' pellets and layers' (or breeder) pellets at the appropriate stages of their growth.

It's best to separate birds once the sexes can be identified, as bullying can occur as the youngsters vie with each other to establish their pecking order. Also, when birds reach sexual maturity the cockerels will attempt to mate the hens and will damage their feathers.

Below: Separate breed pens keep show birds in good condition as they grow.

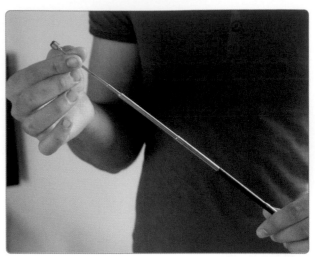

Above: An experienced judge will know if you remove white feathers from a bird.

Above: An extendable magnet doubles as a judging stick.

Having picked a bird that conforms to the standard, you should make sure that it's in good health and has been trained appropriately – it will be sitting in a small cage for hours, surrounded by people looking at it, poking it and generally making a noise, so it's very important to train your exhibit. Pen training, as it's also referred to, basically means confining the bird to a show pen or similar space such as a rabbit hutch for a week or so before the show.

It's beneficial to place this where there's human activity, as this will prepare the birds for the hustle and bustle of a show. Some breeders place the cage on the floor where, if you have one, a dog provides a very active distraction (ensure, though, that the dog can't get at your bird). The potential exhibit needs to be used to this limited space before attending a show, as if a bird is taken directly from a large pen and placed in a show pen for the first time it may become distressed and continuously fly upwards in the pen as it tries to escape, which is both annoying to the exhibitors and alarming to other birds. It may even result in disqualification, as nothing annoys a judge more than an ill-trained bird.

Once the bird has been pen trained it's important to handle it regularly by removing it from the pen and examining it as a judge would, for example by opening up its wings, holding it up to your face and moving the head back and forth, and walking around with the bird in your hands. If you have one, a judging pointer is a useful tool to poke into the chicken's cage every now and then.

Once you've selected and trained your bird(s) for a show make sure that the beak and claws are regularly trimmed. As mentioned earlier you must avoid the quick in both claws and beak, as this will bleed profusely if nicked. Dog nail-clippers or toenail clippers work very well.

Don't modify the bird in any way by, amongst other things, plucking out or dyeing feathers, gluing in extra feathers, or darkening combs with boot polish; a good judge will spot any attempt at fakery and will disqualify the bird.

It's important to keep your birds free from lice, mites and fleas, as the last thing a judge wants is to be crawling with parasites, which will then also spread quickly to neighbouring exhibits. Several good louse powders are widely available, and there's also the Frontline brand used for cats and dogs, although the latter will need to be obtained from a vet by prescription and you may have difficulty getting it even then, as it's not licensed for poultry.

All shows have a closing date and your birds must be entered before the deadline. Show dates are published in various poultry magazines and the local press, which will provide details of the show secretary, who you should contact for a schedule that will list the classes available. Pick the class in which you wish to enter your bird, making sure that it's appropriate – *ie* the correct breed, size, age and sex – and pay the necessary fee.

Below: Birds can be fed and watered when judging has finished.

Left: Well trained exhibits are a pleasure to show.

Confirmation is usually sent before the show with details of classes entered, and when you attend the show your pen numbers will be available from the show secretary, with the slips of paper often set out on a table at the entrance to the show area.

On the day of the show make sure your birds are penned before judging starts (the time will be advised in the schedule). It will help to prepare yourself a checklist the night before and to keep your birds penned inside, with their carry boxes ready and labelled with class and pen number if you have them. Remember to pack drinkers, feed hoppers, a cloth, cotton buds, wet-wipes and kitchen roll, and invest in some small padlocks if you intend leaving the showing area for any length of time, as thefts do occur.

Once at the show and having found your showing pens, sponge off any face dirt with a make-up sponge and apply a little oil or petroleum jelly to the legs and wattles to spruce them up, making sure that it doesn't soil the feathers. Certain breeds will need a wipe down with a silk cloth.

Birds aren't normally fed or watered until judging has finished, as this can be seen as 'marking your exhibit' – that is, making it known to a judge that it's your bird and perhaps thereby influencing their decision. At some shows water containers are supplied, but it's advisable to take your own along with a container of fresh water just in case, since recent legislation says that you *must* provide your bird with water.

When showing certain crested breeds it's important to have a type of water container that will prevent the crest or muffling from getting wet. This can be prevented by using a small container or by placing tape halfway across a larger container to limit access to the water. And keep an eye out for over-zealous stewards who may want to water your exhibit at a crucial point and thereby ruin the feathers.

After judging has taken place there's sometimes a possibility of speaking to the judge to discuss the merits (and faults) of your birds, and this is invaluable and well worth the effort if you can manage it. Difficult as it may be, try not to take any criticism to heart, as the standards that the judges use are there for a purpose, and all the advice you can get will ensure you learn and improve.

When your birds have returned home it's a sensible precaution to apply some form of flea spray or powder. Show birds should also be kept separate from other stock for a week to ensure that nothing contagious has been picked up.

Below: You can prevent crest soiling by taping across the lip of a D cup.

Washing and presentation

Just before the show you'll need to wash and prepare your bird according to breed. Silkies should be washed as near to show time as possible, flat-feathered birds a couple of days before to allow their feathers to settle. Some birds, however, are spoiled by washing and will need a more long-term plan to keep them clean and just a rub-down with a silk cloth on the actual day. If in doubt visit another show beforehand and speak to other exhibitors.

The main thing to remember when washing your bird is not to stress it; you must be patient and allow yourself time to do a thorough and careful job. Get all the tools you need, and make sure they're in easy reach as there's nothing worse than trying to keep a flapping bird in a bath of water while you're struggling to reach the shampoo.

1 Start by trimming the toenails and beak if necessary. Then stand the bird in shallow warm water to soak its feet, which, once immersed for a few minutes, can be washed with baby shampoo and a soft toothbrush, paying particular attention to the individual feathers and scales (depending on breed) along with the nails, as dirt can accumulate under them. It's easier to take the bird out of the water to do this and stand it on a flat surface, the draining board being ideal.

This first stage will acclimatise the bird to the water before the main body-wash stage. It also works well if you just need to clean up legs made ugly by scaly leg mite.

3 Then work some shampoo (either a baby formula or a specialist poultry shampoo) into the wetted feathers until a lather is built up. Make sure that the entire the body has been soaped. After this you should wet the bird's head before rubbing in a small amount of shampoo with your fingers (a 'no tears' formula is ideal here), and once done it can be washed off before rinsing down the body. As you're doing this pay particular attention to the bird itself, as some will try to lie down or their head may fall forward in a faint; if this happens, remove it from the water and allow it to recover its composure before trying again.

2 With the feet cleaned you can turn your attention to the bird itself. Having returned it to the sink, gently pour water over its body

4 Once the bird has been rinsed two or three times to remove all traces of shampoo, wrap it in a thick towel with just its head showing. Never leave the bird unattended on the side of the sink or table, as they can fall off or escape, flapping water everywhere. Once most of the water has been absorbed, gently pat the feathers with a dry towel and get the bird ready for drying with heat.

5 Wipe around the beak with a cotton bud. Any dust and dirt in the nostrils can be gently cleaned out at this point.

6 If you wish to use a hairdryer then put it on the lowest warm setting and on the slowest speed to cause as least stress as possible. If your bird has a creast, then you will need to dry the head first as this will help reduce stress levels, but take care to avoid direct heat to the eyes. Next, you can turn your attention to the underneath of the bird first as this takes the longest time to dry especially in the fluffier breeds, keeping the feathers moving with your fingers, and it is always helpful if you can get another person to help. If the bird raises its wings, do not be fooled into thinking that it is enjoying the experience as this is a reflex action to cool down, telling you that you need to reduce the heat being applied or give the bird a rest for a while. Be aware that excess heat may throw the bird into moult through stress, not what you want at the end of your labours if you are planning on showing, so make sure that any heat applied is warm rather than hot. When the bird is dry, if you must handle it do so gently, as often when it struggles it will mess down itself, which may mean restarting the washing process from the very beginning.

Lightly coating its legs with baby oil at this stage is beneficial, as the warm bird quickly absorbs the oil and it makes the scales more supple and can restore a more youthful appearance. However, make sure you don't get it on the feathers.

So, a few points to remember:
- Never leave a bird alone, either in the water or on the side.
- Don't stress the bird as you wash it. If the bird becomes panicked, stop for a little while.
- Talk to the bird as you wash it, as it will make it and you both feel much better.
- Dry the bird slowly (it may take all day), so leave yourself plenty of time.

Drying box

A drying box is the best option and a must if you have a number of birds to wash, as each compartment will hold a couple of birds and will save a great deal of time, as well as making the process less stressful. All you need to do is place the washed bird in the box, which has been filled with clean wood shavings, place a piece of wire mesh over the top and suspend a heat lamp over the box, high enough to ensure that if the bird jumps up it won't burn its head on the bulb. A distance of at least 30cm (12in) is good to start with, but if the bird begins to show signs of stress or pants raise it another 8cm (3in).

HEALTH PROBLEMS

Common disorders and diseases

Whether you raise birds for the table, for exhibition, as pets or for use in the kitchen garden, knowing that they're in good health is an essential part of your daily routine. But how exactly does the novice, or for that matter the more experienced keeper, diagnose if something is wrong and get the appropriate treatment? The trick is in identifying whether that runny eye is being caused by a feather or a virus, or if that sniffle is being caused by water up the nose after drinking too quickly, or by an infection such as infectious laryngotracheitis.

Many articles and books exist about disease specifics, such as *The Chicken Health Handbook* by Gail Damerow (a US publication, but still useful for UK readers), and *Diseases of Free-range Poultry* by Victoria Roberts. Both of these include really good diagnostic guides and are useful additions to your poultry library, as they've been written by people experienced in keeping poultry.

Disease comes in many forms, and can be caused by a number of different agents which may not in themselves produce symptoms but will leave the bird open to secondary, opportunistic infections which may all have similar symptoms. The main health issues that you're likely to encounter are of three types: parasites, infections (viruses and bacteria), and physical trauma caused by the environment, all of which we'll look at in more detail later in this section.

Some of the most common diseases that affect chickens are detailed opposite.

Above: A partial moult is not uncommon.

> ### Fact...
>
> Despite the name, chickens aren't affected by chickenpox, which is a purely human disease.

Below: Diseases can be caused by environmental factors.

Below: The raised scales of scaly leg.

Name	Common name	Caused by
Aspergillosis	Farmer's lung	Fungi – Aspergillus
Avian influenza (AI)	Bird 'flu	Virus – Orthomyxovirus
Botulism	Limberneck	Toxin
Bumblefoot		Bacteria – Staphylococcus aureus
Cage layer fatigue		Nutritional – calcium deficiency, lack of exercise
Coccidiosis		Protozoan – Eimeria sp.
Egg bound		Environmental – oversized egg
Favus	White comb	Fungi – Microsporum gallinae
Fowl typhoid		Bacteria – Salmonella gallinarum
Gapeworm		Parasite – Syngamus trachea
Histomoniasis	Blackhead	Protozoan – Histomonas meleagridis
Impacted crop		Environmental – improper feeding
Infectious bronchitis	IB	Virus – Corona virus
Infectious bursal disease	Gumboro	Virus – Birnavirus
Infectious coryza	Infectious catarrh, roup	Bacteria – Haemophilus paragallinarum
Infectious laryngotracheitis	ILT	Virus – Herpes virus
Lymphoid leukosis		Virus – Avian leukosis
Marek's disease		Virus – Herpes virus
Moniliasis, sour crop	Thrush	Fungi – Candida albicans
Mycoplasma	Air sac disease	Bacterial organisms – M. gallisepticum and others
Newcastle disease	Avian distemper	Virus – paramyxovirus
Omphalitis	Mushy chick disease	Bacterial infection of the umbilical cord stump
Pasted vent		Environmental
Pasteurella	Fowl cholera	Bacteria – Pasteurella multocida
Prolapse		Environmental
Pullorum	Salmonella	Bacteria – Salmonella pullorum
Red mite		Parasite – Dermanyssus gallinae
Scaly leg		Mite – Knemidocopotes mutans
Toxoplasmosis		Protozoan – Toxoplasma gondii
Trichomonaisis	Canker	Protozoan – Trichomonas gallinae

Internal parasites

Worms

All worms progress through the same developmental stages in their life cycle: mature worms lay their eggs in infected birds which are then passed out in faeces, and once outside the body the eggs undergo development either within an invertebrate host such as molluscs (slugs and snails) or directly in the soil. In the case of a direct life cycle, birds become infected by eating eggs or the hatched larvae from contaminated ground. For those worms with an indirect life cycle, an intermediate host ingests the eggs or hatched larvae. Chickens become infected when they consume the host containing the infective larvae. The length of the life cycle of the various species can last from days to months.

An affected bird will look dishevelled, with its feathers becoming ruffled and dry, and may develop a sharp keel with very little flesh on the breast. The comb and wattles may become pale and the bird may have muck around the vent opening that indicates diarrhoea.

You may also notice a drop in egg production. Weight loss, malnutrition, and depression are typical symptoms of a worm burden in general, and as worms don't always result in death, infestations tend to be chronic in nature. So if you have a large free-ranging flock the chances are that you may not notice the few that are experiencing problems.

There are three major kinds of worm – roundworms, tapeworms and gapeworms.

Below: A mucky bum may indicate illness.

Roundworms (Ascaridia galli)

The most common worms that affect poultry are called roundworms or nematodes, and the most common of these are known as ascarids, which can affect chickens, turkeys, doves, ducks and geese. They can be transmitted either in a species-specific, direct bird-to-bird life cycle, with transmission by ingestion of infective eggs or larvae, or they may have an indirect cycle requiring an intermediate host.

They're usually elongated, cylindrical, un-segmented and large, reaching up to 75mm (3in), and usually live in the intestine, using the chicken's nutrients while the larvae damage the intestinal wall. They can also be found in the oesophagus, crop, gizzard, oviduct or body cavity.

Piperazine, a common wormer, will eliminate roundworms by weakening and paralysing the adult worms, but reinfection can occur through contaminated litter. Piperazine is best given in drinking water as per the directions supplied, repeating at seven to ten days.

The most typical symptom of roundworms is unexplainable weight loss, but they can also cause a drop in egg production as the bird's system tries to cope with the reduced availability of nutrients caused by the feeding worms. Although usually not severe infestation may be particularly debilitating in growers or during the moult.

Tapeworms (cestodes)

Less common but no less of a problem than roundworms, tapeworms are flattened, ribbon-shaped worms with visible segments to their bodies. They require special treatment, but usually don't constitute a hazard to the chicken's health, unless large numbers are present. Reproduction is via segments that break off and are passed through the chicken, contaminating the ground through its droppings. Slugs and snails then pick up the eggs from contaminated vegetation and provide an intermediate host for the larvae, which hatch and migrate from the intestines to the body cavity.

Over a period of approximately three weeks, the eggs form cysts that stay dormant in the host until they're ingested by your chicken. Once inside the intestine of the bird, the actions of digestive juices release the cysts, which attach to the intestines of the bird and new segments form, all within a period of about three weeks. The life cycle is therefore completed in about six weeks.

> ### Fact...
> According to ancient wisdom the best time to treat worms is when the moon is full, which is when they're supposed to be at their most active.

Gapeworms (Syngamus trachea)

Gapeworms normally live in the trachea (throat area) of pheasants, and may be found in chickens kept where there's a high pheasant population. Infestation causes respiratory distress due to damage to the lungs and wind pipe, and is characterised by gaping (opening and closing of the beak), head shaking and neck stretching, sometimes with bubbles around the beak as the birds attempt to dislodge the worms. Tracheal rales (a gurgling sound made during breathing that accompanies tracheal irritation) can be heard in many cases, and can sometimes be mistaken for an upper respiratory infection or some other cause.

Gapeworms use earthworms as an intermediate host, and following ingestion by the chicken the larvae penetrate the wall of the intestine and migrate to the lungs and then to the bronchi. After a larval moult, the adult gapeworms travel up the respiratory system to the windpipe, where the male and female worms intertwine and attach themselves to each other permanently. The entire process from the time the bird ingests the earthworm to the time adult gapeworms can be found in the trachea is approximately seven days. Worm egg production begins about 14 days after infestation of the larvae, with eggs being coughed up into the mouth of the bird, swallowed, and then passed out into its faeces. In the droppings, the eggs incubate for 8 to 14 days under optimum conditions of temperature and moisture to become infective larvae, thus completing the life cycle.

Treatment

Unless you know that a particular area has a heavy infestation of worms, it's unlikely that, if given free-range, a bird will pick up many, and one of the best prevention methods is to regularly rotate your flock to fresh ground. However, as a precaution don't allow an area to become waterlogged, as this encourages snails, which are a host for part of the life cycle. Also, regularly 'rest' areas in order to break the life cycle of the worms.

The use of a licensed product such as Flubenvet (contains flubendazole) twice a year, once in spring and again in the autumn, will control any infestations that your birds may have picked up. This can be bought on the

Above: Flubenvet is licensed for medicating poultry.

Internet if you don't have a supplier nearby, although you may have to sign a POM-VPS declaration (see box).

Incorporate the required amount of Flubenvet into a complete ration of pellets only, for seven consecutive days, meaning that you forego the afternoon grain ration if you normally give it. Mix it into a cup of feed containing a little olive oil. Since the amount is small it needs to be mixed thoroughly for even distribution. Then add this to a bowl of feed and mix again. Repeat twice more, every time into a larger volume of food.

Free-range birds on land with known worm infestations are susceptible to reinfestation, so treatment with another seven-day course after three weeks is recommended.

Note that meat birds treated with Flubenvet mustn't be slaughtered for human consumption until seven days after the last treatment was administered (ie the meat withdrawal time is seven days). However, there's no withdrawal period for eggs produced for human consumption so long as the product has been used at its recommended inclusion rate.

Verm-X, a newer product on the market, is a 100% natural treatment containing garlic, peppermint, thyme, cinnamon, echinacea and other herbs. It's suitable for organic systems and people who don't want to use chemicals to treat their birds. Also, being herbal, there's no need to stop using the eggs whilst treating your hens. It's

Fact...

The acronym POM-VPS ('prescription-only medicine – veterinarian, pharmacist, suitably qualified person') indicates that a medicine is for food-producing animals (including horses), and can be supplied only on a veterinary prescription, which must be provided by a veterinarian, pharmacist or SQP either orally or in writing and must be supplied by one of those groups of people in accordance with the prescription. The term covers what were formerly PML livestock products, MFSX products and a few P products.

Withdrawal periods

A 'withdrawal period' is the legally determined length of time between a bird being given its last dose of medicine, and its meat or eggs being deemed fit for human consumption. There are several important reasons why withdrawal periods are imposed:

- Certain drugs are absorbed into the hen's body, being transferred via the bloodstream to its tissues before being eliminated via the liver and kidneys.
- Drugs are eliminated from the system at different rates.
- Substances can be deposited in eggs ten days before the egg is laid, so if you're using a medication for meat birds it may be prudent to add ten days to any specified withdrawal period.
- Antibiotic residues in meat and eggs may cause a severe reaction in individuals who are allergic to antibiotics. A disturbance of the gut flora may also occur.

available in liquid and pelleted feed forms. If you use the liquid variety, apply one pump action per bird to their drinking water or soaked into bread, repeating for three consecutive days every month and removing all other sources of water to ensure that the birds take the medication. If you use the pellets, add 2.5g (0.09oz) per chicken to their feed for three consecutive days and repeat every four weeks.

Right: Verm-X liquid comes in a handy pump dispenser.

Below: Use the treated pellets as a top dressing on the normal feed.

Verm-X

Below: Verm-X poultry pellets.

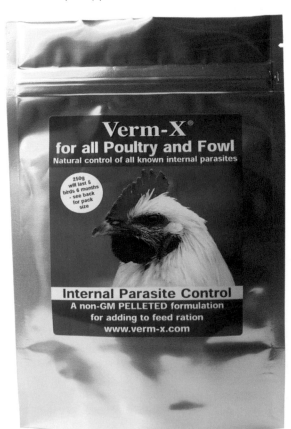

Verm-X

Above: Internal parasites will produce a pale face.

Protozoans

Besides worms, birds may be affected by protozoans. These are minute, single-celled, animal-like organisms that proliferate wherever moisture exists.

Coccidiosis

This is a common disease of poultry caused by the coccidia protozoan, which affects the digestive tract and is primarily found in chickens and turkeys. Despite much research being devoted to advancing the control and treatment of this disease it remains one of the most common in poultry flocks.

Chickens are susceptible to a number of different strains of coccidia, which are host-specific, meaning that they won't affect other species. After an outbreak of a specific species of coccidia, the flock will develop a resistance to that species but remain susceptible to other infective species, meaning that a flock may experience several outbreaks of coccidiosis, each being caused by a different strain.

It usually occurs in growing birds and young adults, and is seldom seen in birds under three weeks old or in mature birds.

PRINCIPAL SYMPTOMS
- Pale combs/faces.
- Ruffled feathers.
- Unthriftiness, lethargy.
- Head drawn back into shoulders.
- A hunched appearance.
- Diarrhoea that may have blood in it.
- Mortality in some cases.
- Tendency to huddle.
- Water and feed consumption drops.
- Laying hens may experience a reduction in rate of egg production.

CAUSES

Coccidiosis is transmitted by direct or indirect contact with droppings via the contaminated drinkers, feeders and litter of infected birds. When a bird ingests the organisms they invade the lining of the intestine, damaging it as they multiply. Within seven days of infection, the coccidia shed immature offspring referred to as oocysts (pronounced oosts) that are then shed by the host in their droppings. The oocysts are not capable of infecting another bird unless they undergo a maturation process in the litter lasting one to three days if the litter is warm and damp, but much longer if the conditions are cool and dry.

The number of infective coccidia consumed by the host is a primary factor in the severity of the resulting infestation, where a small amount may be mild enough to go unnoticed while a large dose may produce severe lesions that can cause death.

Coccidia survive for long periods outside the bird's body and are easily transmitted from one house to another on contaminated boots, clothing, free-flying birds, equipment, feed sacks, insects and rodents.

Above: Overcrowding will encourage disease.

TREATMENT

Separate affected poultry and use a medication for coccidiosis in the water. You may need a veterinary prescription for some of the medicines available. Severely affected birds may need syringing to get the medicated water into them.

Definite diagnosis is determined by the microscopic examination of scrapings of the digestive tract and identification of the coccidia organisms. Since it's common for healthy birds to possess some coccidia, flock history will also be considered by your veterinarian before making diagnosis and treatment recommendations.

PREVENTION

- Keep poultry on a wire floor through which their droppings can fall.
- Adding coccidiostats to the growers' diet can help build up immunity to coccidiosis.
- Give birds free-range as soon as possible.
- Avoid overcrowding.
- Clean housing and brooders thoroughly, preferably with a power hose.
- Don't mix birds of different ages.
- A coccidiosis vaccine is available commercially (brand name Paracox). Seek specialist advice before using it.

Trichomoniasis or canker

This is caused by the Trichomonas gallinae protozoan and primarily infects pigeons, but is passed to chickens mainly through shared and contaminated drinking water. It is a throat and mouth disease that produces a smelly, cheesy discharge.

Above: Chickens will drink from anywhere, so drinker hygiene is important.

PRINCIPAL SYMPTOMS
- Loss of appetite and rapid weight loss with associated weakness.
- Frequent swallowing and an extended neck. An inability to close the mouth or swallow to differing degrees may also be present.
- The mouth may contain sores and/or a foul smelling discharge.

TREATMENT
Move unaffected birds to new clean surroundings and isolate infected birds. Birds not intended for the table can be treated with metronidazole or carnidazole by veterinary prescription.

PREVENTION
- Maintain good hygiene and keep pigeons away from drinking water.
- Avoid mixing new birds.

Blackhead
Although a disease associated mainly with turkeys, chickens can also be affected, although normally to no great extent. Called blackhead because the bird's face darkens, it's caused by the Histomonas meleagridis protozoan and occurs when the chicken eats an earthworm carrying the caecal worm Heterakis gallinarum, in which the parasite lives part of its life cycle.

PRINCIPAL SYMPTOMS
- Ruffled feathers.
- Increased thirst and/or loss of appetite.
- Darkened face, sometimes a bloody discharge.
- Similar symptoms to coccidiosis.

TREATMENT
No effective treatment is available.

PREVENTION
- Improve sanitation.
- Place drinkers and feeders on wire platforms.
- Worm birds regularly to remove the Heterakis worm vector.

Toxoplasmosis
Caused by the protozoan Toxoplasma gondii, this infection is acute (usually fatal) in youngsters and chronic (debilitating but not lethal) in adults. Transmission is via infected droppings, ingestion of an infected host such as an earthworm, or eating infected meat.

SYMPTOMS (OCCUR IN STRESSED BIRDS)
- Lack of co-ordination, circling, twisted neck.
- Pale face, shrivelled comb.
- White droppings.
- Paralysis.
- Blindness.

TREATMENT
None effective.

PREVENTION
- Keep up effective rodent control.
- Practice good hygiene.
- Remove any dead animals from the run immediately.

Below: A rare Narragansett stag and hen.

JANICE HOUGHTON-WALLACE

133

External parasites

These are the organisms that usually feed on the blood or skin of chickens. Attacks may be sudden, especially in the warmer summer months, and severe infestations can cause anaemia and death, so it's important that control measures are incorporated into your daily routine. Such pests are introduced naturally to your flock by wild birds or during visits to shows, where they'll transfer either from neighbouring birds or via your clothing.

Lice and mites come in all shapes and sizes and live by sucking the blood of the bird.

Lice

Menopon gallinae, the common chicken louse, is small (about 2mm or 0.079in long) and yellow/grey in colour. You'll normally find them by parting the feathers under the wings and around the vent/tail area, where they'll quickly scurry out of the light. Their deposits of eggs (nits), which look like clumps of granulated sugar around the base of the feather shafts in the same areas, need to be removed and disposed of somewhere away from your birds. However, they're often difficult to remove, and the feather itself may have to be plucked out completely by grasping it between the thumb and index finger and pulling sharply in the same direction as it's growing. The removed feather should then be burned, nits and all.

While not life-threatening, lice feed off of skin and feather debris and are irritating to the bird, and if present in large enough numbers can reduce its ability to cope with other ailments which may then manifest themselves.

Below: Lice will scuttle out of the light when the feathers are parted.

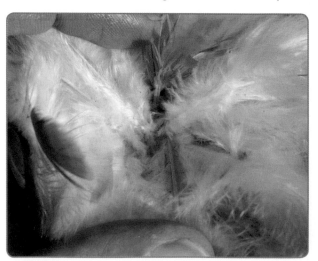

Above: Hold the bird firmly when dusting. Apply the powder and rub in.

The best treatment is with a pyrethrum-based insecticide sold for the purpose, commonly referred to as 'delousing powder'. Lay the chicken on its back and hold it down with your forearm; then, with one hand holding the legs down, the other hand is left free to dust the bird.

Even if you don't see any signs of a lice problem, applying the powder periodically as a preventative measure is advisable.

You may see other preparations recommended on website discussion boards, and these may have been used successfully. However, they're not licensed for poultry (that is, they haven't been tested and certified as being safe to use on chickens) and you'll need to seek the advice of a veterinary surgeon to get them prescribed in order to use them legally. There may also be withdrawal periods associated with them, for both meat and eggs.

Mites

Red Mite (Dermanyssus gallinae)

This mite is about 1mm in size and typically rests in cracks and crevices of the coop during the day before crawling on to the bird at night to feed by sucking blood. Infestations can occur on the bird itself, especially in those breeds with a lot of downy under-feathers, such as Silkies and Orpingtons, and you'll need to take vigorous action to eradicate the problem, since such severe infestations can be lethal.

Most of the time you may not even be aware that the birds are suffering, as you won't normally find any evidence on the bird itself. However, if you look at the ends of the perch, or along crevices in the house, you'll see a pale grey powder residue, and the mites themselves may be present, appearing a deep, blood-red colour when full.

The mites can live for up to six months without feeding, when they pale to a light grey colour. They will then crawl over their unsuspecting host in search of food. In warm weather their life cycle can be completed within just ten days, so you should be constantly vigilant.

Above: Red mite feed on the chicken at night, hiding in the house by day.

Below: Spray the house regularly to prevent mite infestation.

Above: Mite powder is a necessity if you are to stay on top of external parasites.

Scaly leg mite (Cnemidocoptes mutans)

This microscopic mite burrows beneath the scales of the legs, raising them and producing a crust, which is a mixture of the mites' excreta and skin flakes and is incredibly itchy and uncomfortable for the chicken.

PRINCIPAL SYMPTOMS
- Raised leg scales.
- White salt crusts around the legs.
- Discomfort in walking.
- Bleeding on the legs.
- Legs may appear swollen.
- Circulation is impaired and toes may be lost or the foot deformed.

TREATMENT

Soften the crusts for a week or so by applying something like Vaseline, coconut oil or baby oil to the scales (although not too much, as it may get messy). This will help by soothing the irritation, and goes some way towards suffocating the mites. Then a good soak in something like a dog shampoo or a 'no tears' formula shampoo, and a gentle scrub with a soft toothbrush to loosen the deposits, will make the bird feel much better; but if blood shows, stop. Lastly, get some surgical spirit and fill a wide-necked jam jar to the top, and simply dunk each leg in turn up to the hock and leave immersed for 30 seconds, repeating the process every seven days for a period of three weeks.

It may take a while for the legs to look better, often not until the next annual moult, but be assured, they will improve.

Below: Insert the leg into the jar up to the hock.

PRINCIPAL SYMPTOMS
- Birds become depressed; face, comb and wattles pale.
- Egg production is suppressed.
- Small red spots may be present on eggs.
- Birds may scratch.

TREATMENT

Duramitex (contains malathion, now being withdrawn) is a widely used persistent preparation that needs diluting, and is then sprayed into the house, paying particular attention to the perch sockets and any corners and crevices. Preferably you should use a knapsack sprayer. The birds themselves can be dusted with a herbal red mite killer such as Barrier Red Mite Powder. This should be done in the evening, as the red mites feed on birds as they roost at night. Apply in accordance with the product directions, directly on to the birds.

PREVENTION
- Replace felt roofs with corrugated bitumen or clear roofing, to reduce the dark spaces in which mites can accumulate.
- Reduce access by wild birds, as these are the main vector.
- Powder exhibition birds before and after a show.

Above: The ear canal should be examined for signs of mites.

PREVENTION
- Treat the house with a good poultry mite powder.
- Oil the perch with something like vegetable oil.
- As the mites thrive in damp conditions, ensure the house is well ventilated.
- Old-fashioned remedies such as applying creosote or diesel are not to be followed, as these remedies are known carcinogens and harmful to both handler and bird.

Northern fowl mite (Ornithonyssus sylvarum)

A dark brown mite that lives on the body of the bird and is often referred to as 'crest mite' in crested breeds such as Polands. It is similar in appearance to red mite but lives its entire life on the bird and completes its life cycle in less than a week, so infestation is rapid (they can often be seen on the bird in clumps) and if severe can result in anaemia and death.

Crested birds will scratch, and may cause permanent damage to their eyes if a bird is particularly irritated by the mites. Unlike red mite, northern fowl mite actually increases in numbers during cooler periods, which makes them a pest towards the beginning and end of the year when the exhibition season is in full swing. You'll notice the mites crawling across the skin of birds, and they're usually the ones that run up your arms when you handle your chickens.

PRINCIPAL SYMPTOMS
- Birds become depressed; face, comb and wattles pale.
- Egg production is suppressed.
- Bird scratches its head.
- Brown mites are visible.

TREATMENT
The birds should be dusted with a louse killer. This should be applied directly to the birds in accordance with the product's directions for use. With crested breeds in particular a small amount should be applied into the ear canal with a cotton bud, as the mites will hide here and soon multiply. Reapply within five to seven days.

Reinfestation is unfortunately common, but if prescribed by a vet you can use preparations that have a residual effect on the feathers for up to four weeks.

PREVENTION
Reduce access by wild birds, as these are the main vector. Powder exhibition birds before and after a show.

Fleas

Many new keepers mistake lice for fleas, but in fact fleas aren't normally a problem with chickens, and are rarely encountered. A chicken flea does exist, however, and normally inhabits the litter of the house. It is treated by scattering louse powder inside.

Infections

December is the main date in an exhibitor's calendar, and whenever you get a meeting of poultry enthusiasts the conversation always gets round to the subject of ailments; and you can guarantee that whichever bird is being shown, it won't be the best one that an exhibitor has, as invariably their perfect specimen has come down with a mystery illness. Unfortunately this is, nevertheless, the time of year and the place when sniffles and sneezes pass round with lightning speed as thoughtless exhibitors enter birds that are unwell, and by the time the judge has 'passed by' the entry or – worse – picked it up and examined it, the damage has been done, and the disease passed on to the next chicken.

Luckily most owners today keep only a handful of birds at home, and have effectively 'closed' and quarantined their poultry, meaning that you can limit and often escape illness altogether, especially if you manage to exclude wild birds. Even if you do still hear the odd cough or sneeze, try not to be too alarmed, as scaremongering abounds on websites and in some magazines. After all, chickens are as entitled as we are to get the odd cold, and it serves to boost their immune system, which often means that your

Below: The popularity of keeping chickens means that vets are becoming more familiar with their treatment.

pet backyard layer is a lot healthier than some of the top quality pure breeds seen at the shows.

If a cold seems to go on for more than a few days, and especially if you see mucus and/or blood around the beak, and if the eyes are swollen, there are a number of bacteria, viruses or conditions that may have caused the problem, and the only reliable diagnosis will be by a vet. They are now far more knowledgeable than they have ever been, due in part to the growing popularity of poultry keeping.

A tip would be to ask your vet to test for mycoplasma infection, and take with you a copy of relevant literature that highlights this illness, which is more serious than a cold and, while treatable, does leave the bird open to secondary infections and ultimately to a drop in egg production. Antibiotics will usually be prescribed to treat the infection symptoms, and these are quite successful in most cases. One problem that you'll need to bear in mind, though, is that as chickens are a 'prey' species, they do manage to mask symptoms quite well, having evolved over the centuries to keep going for as long as possible in order to prolong their longevity; so by the time you see any clues such as a runny eye or a snotty nose, there's sometimes little you can do to help. Consequently prevention by limiting contact with other birds is definitely the best option.

What other things can you do? Well, if you prefer a natural, preventative approach, then there are natural additives available that can be added to drinking water. Products such as Respite and cider vinegar will improve your chickens' overall health and well-being and make them more robust, whilst a product called Poultry Spice is also useful when mixed in with their feed, and actually attracts them. A balanced diet is essential, along with a stress-free environment. Should you notice a problem, and even if you think you've noticed a problem, isolation and warmth work well in aiding recovery, as do peace and quiet.

Diseases may be bacterial or viral in nature. Bacterial infections, which are typically localised (ie they affect a specific part of the body), respond well to antibiotics as long as the strain isn't resistant, and if caught early enough respond well to treatment. Viral infections, which are typically host-specific and systemic (ie they affect many systems of the body at the same time), are generally not treatable – you have to wait for them to run their course. The secondary infections that follow can be fatal, although as they're normally bacterial in nature they are treatable. Where available, vaccinations for viral infections will protect your stock, and are usually administered at one-day old or as chicks, depending on the virus.

Surviving a bacterial infection may leave a bird in a weakened state, and there's no guarantee that it won't

Above: Always buy in healthy pullets from a reputable seller.

become infected again. Surviving a viral infection, on the other hand, can lead to immunity, although there's often a shedding cycle whereby the infected bird can infect other (healthy) birds, so it's important to keep age groups separate in order to prevent adults that are carriers from infecting chicks at a delicate stage in their development.

Bacterial infections

Salmonella

During the Second World War, the political spin doctors of the time decided that in order to reduce the occurrence of salmonella outbreaks from duck eggs, it would be useful to propagate the myth that duck eggs were poisonous; people therefore ate less of them, and the outbreaks subsided. Nowadays, duck eggs are collected in hygienic conditions, and, like commercial chickens, stock is vaccinated.

There are seven types of salmonella bacteria, only some of which are infectious to humans (zoonotic), and poisoning usually results from having eaten contaminated food such as meat and eggs. Infection causes severe diarrhoea, cramping, abdominal pain, nausea and sometimes vomiting. The symptoms can last for several days, but most people make a full recovery within a week. The symptoms are different in chickens.

PRINCIPAL SYMPTOMS
- Diarrhoea.
- Lethargy.
- Ruffled feathers.
- Drooping wings.

TREATMENT
There is none available. An infected flock will need to be culled and the area disinfected thoroughly, as birds that recover remain carriers thereafter and shed the organism throughout their lives, infecting others as they go.

PREVENTION
- Transmission can also occur through eggs, so it is recommended that you wash hatching eggs in a warm solution of Virkon S, an egg sanitising powder that's soluble in water.
- Buy in vaccinated stock.
- Rats, mice and wild birds can also spread the disease, so hygiene and pest control are important.

The European Commission has set targets for reducing salmonella in laying flocks and making vaccination compulsory in countries with high contamination levels. Commercial flocks in the UK are vaccinated, so if you buy your stock from a registered hatchery the parents will have been injected with a vaccine to provide a high level of antibodies in their chicks.

Although vaccines are also available to the hobbyist (indeed, DEFRA are aware of the popularity of backyard flocks), they'll need to be administered by a vet, and the

Above: Most backyard flocks are unvaccinated.

cost is often prohibitive to people keeping just a few hens, as their birds have to be examined beforehand to make sure they're clear of infection. Unfortunately many of our backyard flocks are therefore unvaccinated and may well be infected, and most of the time you'll never know that your birds are carrying the bacteria. As a precaution, ensure that the elderly, the young, pregnant women and those with a compromised immune system cook eggs thoroughly to reduce any risk of food poisoning.

If you suspect that your flock may be infected specialist diagnosis is vital, and you'll need to call your veterinary surgeon to arrange for blood and faecal sample tests.

More information can be found on the DEFRA website at http://www.DEFRA.gov.uk/animalh/diseases/zoonoses. Remember that since some salmonella infections can transfer between animals and humans you're just as likely to pass it to your flock as vice versa, so always wash your hands before you handle your poultry as well as after.

The main types of salmonella affecting poultry are:

SALMONELLA PULLORUM
BWD (bacillary white diarrhoea) is passed on through infected eggs, but is easily spotted with a blood test. It's specific to chickens, turkeys and pheasants. Symptoms (which are mainly confined to birds in their first month of life) include many chicks dead in their shells or dying shortly after hatching, while older birds may have white faeces stuck to the vent feathers. Treatment is best carried out using antibiotics from your vet, although treated birds may continue to be carriers.

SALMONELLA GALLINARUM (FOWL TYPHOID)
This is specific to all poultry and is diagnosed by blood test. Symptoms include yellow diarrhoea, suppressed appetite, pale comb and wattles, ruffled feathers and a general lack of condition. Treatment is best carried out using antibiotics from your vet, although treated birds may continue to be carriers.

SALMONELLA TYPHIMURIUM AND SALMONELLA ENTERITIDIS
These are the zoonotic variants affecting chicks, which will exhibit pasted-up vents, a strong smell, ruffled feathers and a high death rate.

Mycoplasma
Sometimes referred to as 'roup' in older poultry publications (as are a number of similar conditions), there are a number of mycoplasma bacterial infections, with the main two affecting either the respiratory system (m. gallisepticum) or the joints (m. synoviae).

PRINCIPAL SYMPTOMS
- M. gallisepticum is first seen as a foam in the corner of the eye.
- This is later followed by rattles in the chest, swollen sinuses and general respiratory distress.
- The infection will often debilitate the bird sufficiently that other infections, either already present or passed on, will proliferate.
- M. synoviae affects the joints, making the bird lame.

Above: Foam in the corner of the eye.

TREATMENT

This is best carried out with antibiotics from your veterinary surgeon. Keep the bird warm and quarantined, and wash away any accumulations of mucus.

PREVENTION

- Mycoplasmas are very contagious and transmitted easily, either by wild birds or on your boots and clothing when returning from a show, as it stays live for several hours. Good hygiene and steps to prevent wild birds coming into contact with your chickens will help.
- Immediate treatment at the first signs of disease will limit its effects, although treated birds will remain carriers.

Right: The infected scab typical of bumblefoot.

Bumblefoot

Caused by the staphylococcus bacterium that's found normally on the skin. Entry is via a wound as the bird lands and damages the pad of the foot. High perches have been blamed, but some breeds may be predisposed to the infection, and it may be passed on through the egg. The infected foot will fill with pus and become inflamed and painful, and in severe cases will rupture through the top of the foot.

SYMPTOMS

- Lameness.
- Inflamed, hot foot.
- Abscess with a black scab may form on the pad of the foot.

TREATMENT

It's best to consult your vet unless you're experienced, as you'll need to administer oral or injected antibiotics to treat the infection. If you're going to treat the condition yourself, then you'll need to begin by soaking the foot in an antiseptic solution to soften the plug and reduce the chance of further damage and additional infection. Wear gloves at all times, as the staphylococcus bacteria can also cause infection in humans.

Remove the plug of pus, and clean the hole out with an antiseptic wash and pack it with a sulphur product. Spray the cleaned area with an antiseptic wound wash and securely bandage the area. Burn any residual matter.

PREVENTION

- Good sanitation is essential.
- Ensure perches are smooth and rounded and not too far from the floor.
- Don't breed from parents that have suffered from bumblefoot, as it may be transmissible through eggs.

Gumboro (infectious bursal disease)

This normally affects youngsters of around three to six weeks and attacks the bursa of Fabricius (see page 164). The sudden onset will kill a chick within two days, and if it lives it will be permanently susceptible to secondary infections.

SYMPTOMS
- Hunching.
- No interest in food and water.
- Depression.
- Similar in appearance to severe coccidiosis.

TREATMENT
A highly infectious disease, treatment with antibiotics will keep deaths to a minimum.

PREVENTION
There is a vaccine, which works best when given to breeding hens that will pass the immunity to the chicks.

Botulism (limberneck)

Caused by a bacterium found in rotting organic matter, this is ingested by pecking at degrading plant and animal matter.

SYMPTOMS
- Paralysis, starting at the feet and moving upwards.
- Floppy neck.
- Drooping wings.

TREATMENT
Severe cases are untreatable, but if caught in time an antitoxin may be available from your vet, and often oral therapy with activated charcoal and Epsom salts may be effective to flush the bacterium from the crop.

PREVENTION
- Maintain good hygiene.
- Don't feed your birds rotting meat or plant material.
- Keep drinking water clean.
- Remove any dead animals from runs immediately.
- Control flies.

Infectious coryza

A rapidly spreading respiratory disease affecting all ages. Highly contagious, it can be spread from bird to bird and via contaminated clothing and equipment.

SYMPTOMS
- Watery, swollen eyes, often glued shut with mucus.
- Swollen face and sinus.
- Smelly discharge from the nostrils.
- Diarrhoea.

TREATMENT
Treatment with antibiotics will keep deaths in a flock to a minimum. However, survivors are carriers, so an infected flock is usually culled to prevent reoccurrence.

PREVENTION
- Don't mix new birds in with a flock without a suitable (two- to three-week) quarantine period.
- Don't mix birds of different ages, as mature carriers will infect the youngsters.
- Disinfect housing after an outbreak and leave empty for three weeks.

Omphalitis (mushy chick disease)

A disease of chicks, this infection enters through the unhealed navel and can also be the cause of late death in chick embryos. Transmission is through droppings on eggs and poor incubator and brooder hygiene, as well as the humidity in an incubator being too high.

SYMPTOMS
- Embryos dead in shell.
- Unhealed navel on chicks, often wet or scabby.
- Puffed up appearance, lack of interest in food or water, a droopy head.

TREATMENT
None effective. Cull affected chicks.

PREVENTION
- Only hatch clean, un-cracked eggs.
- Practise good hygiene during incubation and brooding.
- Control incubator humidity.

Pasteurella (fowl cholera)

A bacterial infection known to be transmitted by rats.

SYMPTOMS
- Swollen face, especially the wattles.
- Respiratory distress (gasping, wheezing, rattling).
- Lethargy.
- Lameness.

TREATMENT
Antibiotics are effective and a vaccine is available from your vet.

PREVENTION
- Keep up effective rodent control.
- Don't mix birds of different ages, as adults may be carriers.

Viral infections

Avian influenza ('bird 'flu')

A viral infection of birds, avian influenza (shortened to AI) is found throughout the UK, especially along wild bird migratory routes, and is common in the East of England. Like any 'flu virus it spreads easily from host to host, mutating as it goes, and the main worry is that it will mutate and cross over to human hosts. A zoonotic disease, it is therefore notifiable.

Of all the poultry diseases, this is probably the most alarming to chicken keepers, often unnecessarily so. Many poultry-related websites contain excellent information (www.practicalpoultry.co.uk is one of the best), and it's well worth reading up on the subject to ensure that you're fully aware of the issues and able to separate fact from fiction.

SYMPTOMS
- A difficult infection to diagnose visually, the symptoms are common to many ailments and include:
- Sudden death.
- Cessation of egg laying.
- Respiratory distress (gasping, wheezing, rattling).
- Swelling of the sinus and face.
- Facial haemorrhaging (noticeable in the comb, face and wattles).
- Cyanosis of the comb (a bluish/purple tinge).
- Loss of appetite and weight loss.

TREATMENT
A highly infectious disease, there's no treatment and an affected flock can only be culled. If you suspect AI then you must consult your vet immediately and advise DEFRA.

PREVENTION
- A vaccine is available which may be used in the event of an outbreak.
- Isolate chickens from wild birds.
- Implement strict biosecurity measures (see page 83).

Vent gleet
This is a disease caused by a herpes virus and is passed on through mating.

SYMPTOMS
- Yellow, smelly moist covering around the vent.
- The tissue may be necrotic (dead).

TREATMENT
Acyclovir, the treatment for human cold sores, is known to be effective in reducing the symptoms, but the virus will remain in the bird and periodically resurface.

PREVENTION
- Do not mate from infected birds.

Infectious bronchitis (IB)
Since it causes respiratory and kidney disease in younger birds, and reduces egg-laying capacity/quality in older birds, this is a significant disease in the commercial layer industry. If eggs are laid they're often wrinkled in appearance.

The infection can travel through a flock in as little as one to three days, while mycoplasma, which shows similar

Above: The vent should be examined for signs of muck, lice or infection.

symptoms, is slower and affects fewer birds. Since there's the possibility of infectious bronchitis also being confused with Newcastle disease and infectious laryngotracheitis, laboratory diagnosis is essential.

SYMPTOMS
- Gasping, coughing, sneezing, rattling.
- Water discharge from the nostrils.
- Swollen sinuses.
- Drop in egg production.
- Increase in egg deformities, including runny egg whites.

TREATMENT
Keep infected birds warm and well fed and give them glucose in their water. A broad spectrum antibiotic will treat any secondary infections.

PREVENTION
- Control mycoplasma infections to reduce the chance of IB taking a firm hold on a bird's system.
- Vaccination for chicks is available.

Infectious laryngotracheitis (avian diphtheria, ILT)
This is a herpes-type virus that mainly affects older male birds, although all ages and sexes can be affected. The main symptoms are respiratory distress with plugs of mucus building up in the throat. Infected birds may cough up or sneeze blood and the mortality rate is high. The whole flock can become infected, normally within two to six weeks, as ILT is spread easily by bird to bird contact, contaminated people and equipment, and poor manure disposal. Mycoplasma and IB infections increase the likelihood of death.

Above: A twisted neck is indicative of many conditions.

PRINCIPAL SYMPTOMS

Mild infections:
- Watery inflamed eyes.
- Swollen sinuses.
- Nasal discharge.

Severe infections:
- Watery eyes.
- Nasal discharge.
- Coughing up blood.
- Gasping, gurgling, rattling, 'cawing'.
- Swollen sinuses.
- Depression.
- Depressed food intake.

TREATMENT

Antibiotics will treat any secondary symptoms. A vaccine is available and is dropped into the eye, giving mild symptoms. Disinfect all housing and leave empty for six weeks, as the virus doesn't live for long outside of the bird.

PREVENTION
- Control other infections and keep stocking densities low to limit spread.

Marek's disease

Another herpes-type virus, first identified by Josef Marek. It infects the lymphoid tissue, commonly causing tumours and damage to the peripheral nerves resulting in paralysis, typically of one wing and one leg. Torticollis may be

Above: A twisted neck is indicative of many conditions.

present but should not be confused with water on the brain. Marek's can affect most of the systems of the body where tumours interfere with normal functioning, eg lesions and tumours in blood vessels and the digestive system.

Infection is by the inhalation of feather debris, as this is where the virus sheds. The virus quickly develops in the body and becomes latent until symptoms appear between four weeks and several months later, usually as a result of stress. Certain breeds such as Silkies and Sebrights are more susceptible, with females more so than males.

Torticollis

Commonly referred to as twisted neck, wry neck or seahorse neck, torticollis is an outward symptom of a problem affecting the nervous system, such as an infection which – in order to reduce the tissue swelling that's pressing on the nerve causing the neck to twist – will need to be treated with antibiotics and an anti-inflammatory. Rest and warmth will calm the bird and reduce the severity of the muscle spasm bending the neck, especially if a stress-related condition has caused the secondary infection that's led to it.

SYMPTOMS

- Tumours.
- Paralysis, with one wing and one leg sticking out in opposite directions.
- Drop in egg production.
- Increased moulting.
- Torticollis.
- Disorientation.

TREATMENT

Antibiotics may reduce certain symptoms, such as disorientation, but birds will most likely die or at best become carriers, shedding the virus. Culling is the most effective treatment.

PREVENTION

- Vaccination at one-day old and again at two weeks.
- Don't keep older birds with younger, as the youngsters will pick up any shed virus. The older a bird is before exposure, the greater its chances of resisting, as chicks will develop natural resistance from about five months of age, helping them to overcome later infection.
- Turkeys can carry a virus that stops the Marek's virus from producing tumours and keepers will sometimes run their stock with turkeys in order for their birds to pick up this virus. However, mixing species is not recommended.
- Breed for immunity.

Newcastle disease (paramyxovirus, fowl pest)

First isolated in 1926 in Newcastle, the UK is mostly clear of this virus, but as it is a zoonotic disease causing 'flu-like symptoms and conjunctivitis in humans, it is notifiable to DEFRA (see notifiable diseases, page 163).

SYMPTOMS

- Respiratory problems.
- Green diarrhoea.
- Drop in egg production.
- Soft-shelled eggs.
- Torticollis.
- Sudden death.

TREATMENT

Infected birds are culled.

PREVENTION

- A vaccine is available but should only be used in the event of an outbreak.

Lymphoid leukosis

Caused by a retrovirus, this disease normally affects birds over six months of age, and appears as tumours throughout the body, especially the liver. Infected chicks will pass the disease on to older birds and other chicks but rarely exhibit symptoms themselves.

SYMPTOMS

- Sudden death.
- Drop in egg production.
- Distended abdomen.
- Tumours may be felt under the skin.
- Cyanosed comb (a blue/purple tinge).
- Shrivelled comb.

TREATMENT

None.

PREVENTION

- Don't mix birds of different ages, chicks are a source of transmission.
- Keep strict hygiene.
- Disinfect housing after an outbreak.

Below: Infected chicks will pass diseases onto other birds.

Environmental and physical issues

These are by far the most common form of disease that you're likely to encounter, and most are easily treatable and preventable from home.

Feather pecking

Chickens tend to peck each other, since this is, after all, part of the process of establishing the 'pecking order' amongst them, where the highest-ranking bird will peck all those beneath her, while the lowest ranking bird is pecked by all. Normally this isn't a problem, as you'll notice that the bird being pecked simply runs away. However, sometimes this isn't possible, and there may be other factors involved. So try to identify the cause of the pecking. Number one on the list of causes is boredom, especially in enclosed systems where there's nothing to do; but this is also the easiest to cure,

Below: Suspended nettles will provide entertainment and food.

simply by giving the birds something to occupy them, such as hanging up a bunch of greens, which makes them jump up.

Stress is also a factor and, surprisingly, can be caused by glaring or harsh white light. If you think this is the problem, then simply changing to daylight bulbs or a lower wattage will help. Heat will likewise cause stress, in which case you could try increasing the ventilation. Remember too that chickens are attracted to red (which is why the base of some feeders is red), so if a feather is broken – and this is especially valid if new feathers are growing through after the moult, or with chicks – it will attract further attention which can escalate until a fatality occurs. In this instance, a squirt with gentian violet spray (which should be a staple of your first aid kit, both to disinfect and to discolour) plus a squirt of 'anti peck' spray will work wonders.

Above all you should act promptly and without hesitation. Birds that constantly peck at others – which you should be able to distinguish from behavioural pecking, as it will be sustained, and may be aimed at a particular area of the body – should be immediately isolated in a separate (temporary) run next to the main run, during which time you allow the other birds to maintain contact so that when the dominant hen is reintroduced the victimised bird isn't attacked again.

Isolation of the dominant bird should continue for approximately three to four days to give her time to forget about her victim. Meanwhile if the pecked hen has been injured she should also be isolated until fully healed, in order to protect her from cannibalism. Reintroductions should be in the evening as the birds begin to roost.

Moulting

As we've seen, for most birds the moult is an annual process that lasts for up to two months, but you'll often find one bird in a continual state of moult, which may indicate an underlying condition. Moulting also weakens a chicken's resistance to disease, as the immune system is compromised and any suppressed conditions are able to flourish.

The barring of feathers may become apparent in some breeds when there's a nutritional problem – for example, a black feather with a green sheen may develop purple barring.

Aggression

In males

Your solution to this problem lies in understanding why a male chicken acts so aggressively in the first place, strutting around like the lord of the manor. It's because in effect that's what he is – he's guarding his harem and looking out for contenders to his throne. If you approach him, invariably his hackles raise and his wings elevate slightly, and just before he attacks he'll jump at you sideways, often with a guttural squawk, and slash with his spurs – which can hurt if the bird is large.

Some keepers think that the best way to stop this is to plan your own offensive and mimic his behaviour by jumping heavily sideways, squawking, and raising your elbows, and then – with the advantage of being able to think ahead – pre-empt his attack by either sidestepping or by placing your foot in his way and then launching him skywards when he lands. Although this may work for some, however, retaliation will often just encourage him to make further attacks.

A more effective and recommended approach which works well is to clamp your gloved hand over him when he's close (this can be done easily with practice), flatten him down at the shoulders to prompt the reflex hunch, and then pick him up; this confuses him and diffuses his aggression, and if you walk around with him tucked under your arm he will soon calm down. However, do remember to wear a good thick shirt, as often a quick peck will draw blood from exposed skin.

Above: Placing your hand firmly on the back of an aggressive cock bird will calm him.

In females

This usually comes from the dominant hen, and once she's asserted her authority the problem goes away. New males, when introduced to a group of females, may be attacked, and you'll need to be on hand to ensure this doesn't happen, as a bullied male will not mate. You can pick up an aggressive hen in the same way as the male.

Below: Hens will fight to restore the pecking order.

Boredom

This easily becomes a problem in intensive systems and manifests as feather pecking and vent pecking in adults and toe pecking in chicks. If blood is drawn the birds can quickly become frenzied in their attacks, as they're drawn to the colour red, and cannibalism can result. If you can, keep stocking densities as low as possible, and if you must restrict birds to houses or indoor units, hang up bunches of green stuff to provide a distraction.

Wounded birds should be treated with gentian violet spray to act as an antiseptic and to mask the red with purple, a less attractive colour. Remove the bird if necessary and confine to a run in full view of the others.

Boredom will also cause stress, which may manifest itself in various illnesses.

Egg eating

Once a hen has tasted an egg she soon gets to enjoy the flavour, and you're stuck with a vice that's very hard to break, with the culprit – who can easily be identified by the yolk around her beak – pecking at freshly laid eggs whenever she can and consuming the contents. This habit, once started is hard to stop, and prevention is much better than cure:

- Collect eggs regularly, sometimes a couple of times a day in full season, which will prevent them getting trodden on and broken, spilling the contents for an easy meal.
- Having several nest boxes will stop hens developing a favourite and crowding it, which inevitably leads to breakage.

Below: Mustard is often inserted into the empty shell before taping back together with masking tape.

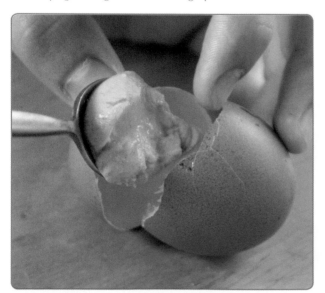

- Ensure the laying area is darkened and quiet, which discourages greedy hens from raiding the nest boxes, and if you can raise them slightly more than standing height so much the better. Egg boxes should be situated at the point furthest away from the door or window. The entrance can be covered with a curtain of plastic to darken the inside further.

You may have some success with mustard and chilli powder put into blown eggs, although a much more effective option is to put china (crock) eggs or ping-pong balls into the nest, which the hens soon tire of pecking as they're not rewarded with a meal. Blowing an egg can be done by placing masking tape over each end to stop cracking and then picking out holes, through which you can then blow the contents, which is very messy and time-consuming. The empty egg can then be filled with your concoction and the hole taped over.

Deformed eggs

Sometimes you may encounter abnormal eggs: soft shells, slab-sided and chalky shells to name a few. If this is infrequent then it's nothing to be unduly worried about, in fact new layers and birds coming off lay may lay a soft shell now and again. If, however, it happens regularly it could be a sign of calcium deficiency, which may mean you need to feed additional calcium grit (sold as oyster shell grit) or, if you've not already done so, switch to a quality layers' pellet feed.

If this doesn't solve the problem then your next step may be to get the birds' blood and poo tested at the vets. But talk

Below: This is all that is left of a soft shelled egg.

Above: An egg within an egg indicates a blip in production.

to your veterinarian first, as more and more people are keeping chickens and vets are consequently becoming more knowledgeable in their diagnoses.

Sometimes disease will result in frequent soft-shelled eggs, and if you're breeding for replacement birds then the hens responsible are best removed from your breeding pens, as they may pass the problem on to their progeny.

When a pullet first starts to lay, her system must undergo a rapid transformation and this results in oddities; similarly, as she gets older shell quality deteriorates and becomes more papery, lighter in colour, and sometimes misshapen. Oddities occur in a number of forms. Probably the most common is an egg with a thicker band of shell round its middle. This is often caused by a shock, such as a loud noise or even a sudden rain storm, when the egg stops momentarily in its passage through the hen's system and extra shell is deposited, resulting in the thickened band and slight elongation. Such eggs are not good for hatching but are fine as far as eating goes.

Slab-sided eggs (where one side is slightly flattened), are less common and are often a sign that all is not right with the hen's system, whether due to a passing cold, old age or something more serious. If your bird doesn't have signs of illness then don't panic, as it may just be one of those things.

Wind eggs and double-yolkers are two extremes of the same problem: the first is a tiny egg with nothing but albumen (the white), where no yolk has been released; the second occurs when too many yolks have been released at once. Neither type will be hatchable if incubated.

Not laying

If you suspect that your hen isn't laying, look at her wattles: if they're bright red then the chances are that she's in lay. Feel, too, for the gap between her pelvic bones: if you can place one or two fingers between them then it's likely she's not in lay; if you can place two to three fingers between them then it's likely she is laying.

Check too that the bird is female, which isn't as obvious as it sounds, as unscrupulous dealers or those who simply lack sufficient knowledge to sex a bird correctly may have sold male birds on.

One of the most common causes for not laying amongst backyard birds, where only a few are kept for eggs, is obesity. Mostly chickens prefer a diet rich in protein and fat, and will always pick the tastiest of morsels (you can test for yourself by throwing in a handful of mixed corn into their feed, as it will always be the maize that's picked out first, which is another reason for advocating whole wheat only). As a result, a steady feeding regime will mean more and more is eaten, and soon your previously svelte pullet becomes a thundering mass of feathers, especially if confined to a run. Hens will then stop laying, while cocks become infertile.

If you want to check if your bird is too fat, just pick her up. A healthy bird will be well fleshed and her breastbone won't feel sharp, but you should still be able to feel it. Slimming down your hen is quite simple. Either limit her intake or switch to mash to make her work for her meal by scratching. In addition hang up a bunch of vegetation so that she'll have to jump for it.

Reduced day length is another cause of not laying, especially as days begin to draw in, because the reduced

Below: Two to three fingers placed between the points of the pelvic bones indicates that this pullet is in lay.

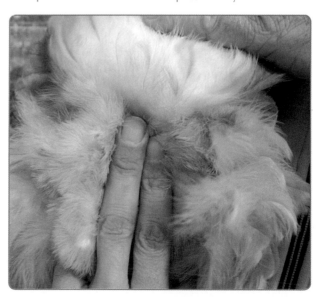

Above: Support the hen with her head towards your body to examine the vent area.

Below: An electric timer can be preset to turn on equipment.

Cage layer fatigue

Cage layer fatigue, or caged layer syndrome, is not a common condition and mainly affects caged production birds of strains specifically bred for mass egg production. Characterised by weakness, squatting and an inability to stand, eggshells become paper-thin and bones brittle due to the bird's inability to move calcium around its body correctly; instead, it draws it from its bones.

The disorder may be both nutritional and environmental, where stress caused by the bird's living conditions upsets the body's system. A milder form is seen in exhibition birds, which can become weak (although the bones don't become brittle) as a result of excess showing.

If you have ex-battery hens that you think may be suffering from this condition, allow them access to sunshine and provide them with oyster shell grit.

amount of light will mean that egg production will drop, a response known as photoperiodic. Indeed, some of the older birds and traditional breeds may stop altogether, which is why historically ducks were used to provide eggs through the winter. Although many keepers are not concerned about this drop off, if you keep your birds solely for eggs then you may want to supply artificial light to keep them going, or have a fresh supply of pullets ready to take over.

Typically extra light is provided at the start of the day so that the birds are still able to roost at dusk. If you intend to do this you may want to invest in a timer to make sure that you don't have to get up yourself, although it will encourage any cockerels you have to crow earlier. You can buy some useful gadgets that sense the light level and adjust the timer as necessary, effectively learning when to turn on the light and making your job even easier.

Egg bound

Normally the egg travels down the oviduct pointed end first and turns over at the end of the journey, to be passed through the cloaca blunt end first. If the egg is the wrong way round, or too large possibly as a result of a calcium imbalance caused by stress (such as in young hens just coming into lay), it will become stuck and subsequent eggs will back up. The hen will stand hunched and may exhibit signs of distress as she strains to pass the egg, which is often made more difficult if the muscles that would normally push the egg have become damaged. In this case the best method of treatment is to keep her warm and quiet and put a little KY jelly or liquid paraffin into the vent with your finger. (Don't do as some people recommend and hold the bird over a steaming saucepan of water, as this will only stress the bird further, or worse.) With the other hand, press gently on her belly to push the egg towards the vent, and you should see it start to emerge. If it remains stuck due to its size, then gently pierce it and remove all the fragments (any left behind can cause further injury). Return the hen to warmth and quiet after you've finished.

Calcium supplements (such as Calcivet, which is water soluble) can be used as a preventative, but aren't recommended long-term as they can interfere with the natural regulation of calcium in the body, causing further problems.

Egg binding may be inherited. It is often seen in heavily inbred exhibition strains, and may be aggravated by showing the bird.

Infertility

This can be due to a number of factors, including stress, lack of available or viable sperm (the sperm of certain rosecomb breeds doesn't live as long in the hen), illness causing lack of vigour and energy, and heavy mite or louse infestations where the immune response of both males and females will be reduced. Some of the fluffier breeds may have excess feathers around the vent area, and these will need trimming to allow mating to be successful. As already mentioned, the moult will also cause temporary infertility. Sometimes an aggressive cockerel or dominant hen will push the mating couple apart, although this is easily solved by removing the bird responsible.

If you see a female with bare sides and possibly flank wounds (see below) it's likely that she's a cockerel's favourite, and by removing her to recover the male will be encouraged to mate with the other hens in the pen.

Below: A saddle fits over the back to protect the bird.

Flank wounds

Though such injuries are often caused by fighting they more usually result from mating, when the cock's spurs scrape off the hen's feathers and eventually damage her skin. Prevention is achieved by regular trimming of sharp spurs and by fitting a poultry saddle if necessary. If a saddle is fitted over a wound, check under it regularly to ensure that infection hasn't set in and that no pest infestation is evident.

Crossed beak

This deformity is usually genetic in nature but can be caused by injury at a young age, and may not be noticeable in the chick but only becomes apparent as it gets older. The upper part of the beak crosses the lower and in severe cases prevents the bird from feeding properly, which may result in undernourishment. Mild cases can be treated by regular trimming of both the upper and lower beak to improve the closeness of fit.

Above: A crossed beak will cause problems when feeding.

Below: Trimming top and bottom will even out the beak, but will not cure the problem.

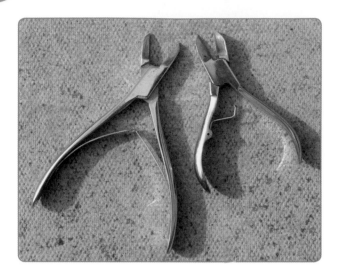

Above: Heavy duty nail cutter will trim beaks and toe nails adequately.

Overgrowth (beak, claws, spurs)

The chicken's beak, claws and spurs grow continuously and are normally worn down by the scratching around involved in foraging. Overgrowth consequently occurs more often in birds without access to free-range. Toenails and beak can be clipped with a sturdy pair of toenail clippers if necessary, while spurs can be blunted with a strong pair of cutters.

As mentioned earlier, the beak and spurs will both bleed

easily if cut too close to the quick. This can be seen as a pale area under the cuticle of both, almost resembling a claw within the claw, or a beak within the beak, although it's difficult to see in darker-coloured birds. Spurs contain a large artery, so care is needed when trimming these; if in any doubt get your veterinarian or an experienced keeper to show you how it should be done.

Splay leg

Affecting chicks in the brooder, or seen at hatching, splay leg is quite a common problem and is caused when the ligaments and tendons in one or both legs aren't strong enough to support the chick and it scrabbles round in circles rather than being able to stand up. Brooders should have a slip-proof surface such as a tea towel or wire bottom to allow the chicks to stand and strengthen their legs. Wire is ideal, as it allows the chick to purchase a grip by wrapping its toes around and push itself up. Some breeders have tried tying loose wool around the legs to bring them together, but this is unreliable and fiddly.

Below: The brooder floor should be slip proof.

Pasted vent

Very common in chicks, this is probably caused by nutritional deficiency as a result of chilling. Symptoms include droppings adhering to the vent, along with a distended abdomen as the faeces accumulate, and a droopy, hunched stance, often huddled under the heat source. Gently remove any droppings sticking to the vent area and wash with an antiseptic solution. By keeping the chicks constantly warm and avoiding draughts you're likely to avoid this ailment arising, and it may help to add a probiotic to their drinking water, which should not be chilled. As it's not really understood why it occurs, don't automatically assume you've done something wrong if you encounter it.

Lameness

This can be caused by a number of factors and can be inherent in broilers, where birds are bred for rapid growth in enclosed systems and their bones aren't given time to mature.

Many infections, such as mycoplasma, can settle in parts of the leg, including the tendons, joints and bones, and can be treated with antibiotics when prescribed by your veterinarian. In poultry, kidney disease can also cause lameness, while in waterfowl a heavy worm burden can result in the bird going 'off its legs'.

Hypermetria is a high-stepping gait caused by a malfunction in the cerebellum (part of the brain) and has no cure, although affected birds manage well.

Deformities

These are normally either inherited conditions or caused by early injury, and unless the condition isn't causing undue stress and you don't intend to breed culling is the best option. Conditions include roach back (backbone bends upwards), wry tail (tail bends to one side), cow hocks (legs bow inwards), bent toes and bent breastbone.

Cuts

These are best treated with veterinary wound powder and the bird isolated for a while until recovered. Gentian violet spray is a great natural antiseptic coverall.

Breaks

These are rare in chickens that are being fed properly, but can be splinted if they occur.

Moniliasis (sour crop)

This condition caused by candida yeast is also known as thrush, and can occur, as in humans, as a result of antibiotic treatment. It can be recognised by a powerful sour smell from the crop and a lethargic stance. Treatment is usually with nystatin, which is a strong antifungal drug of bacterial origin.

Impacted crop

Also known as crop binding, this condition is caused by a blockage of the crop and is common in birds ranging on long grass. When squeezed gently the crop feels like soft plasticine, and the best treatment if caught early enough is to soften the mass by pouring a little warmed olive oil or liquid paraffin into the throat, best administered by means of a large syringe. The bird is then suspended upside down and the contents of the crop are massaged out towards the mouth, allowing regular breaks to prevent suffocation.

However, if you're in any doubt at all don't attempt to do this, as you'll cause more damage and may kill the bird; it would be better to seek veterinary advice, and the vet may well open up the crop to clean it out. Birds treated in this manner do recover but are prone to further bouts.

Below: Syringing is easier with two people.

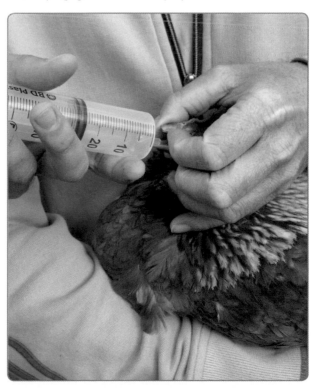

Prolapse

This is a particularly frightening condition when first encountered by a keeper, as the bird's insides hang outside its body as a result of the connective tissue that normally holds the organs in place being damaged. Unfortunately the first you may know of it is a dead hen with her rear end pecked, as her flock mates are attracted to the red flesh. If you find the bird alive, immediately isolate her and keep her warm and quiet. You can attempt to put everything back in by gently turning the bird so that her head is pointing downwards and applying a little warmed liquid paraffin to the prolapsed organs. Then carefully manipulate the insides back towards the vent, and if she and you are lucky gravity and the body's natural rhythms will take everything back in, especially if it's only a mild prolapse and you have haemorrhoid cream to hand, which may reduce the swelling. Larger prolapses will need veterinary treatment, and you should ask for antibiotics in all cases of prolapse, to combat infection.

Birds that recover may prolapse again when the next egg passes through, or, indeed, may never lay again. The problem may be inherited in certain pure bred exhibition lines.

Pendulous crop

Caused when the muscles in the crop weaken and are no longer able to push food through to the stomach, this disorder is characterised by the permanently distended crop hanging below its normal position in a bird's chest. It may also swing as the bird walks. Don't get this confused with a full crop at night, which looks similar but is gone by morning. The best treatment is to isolate the bird and give it just water for 48 hours, and then gradually reintroduce the normal feed ration.

Below: The crop should be firm but not hard.

Above: Apply petroleum jelly if severe cold threatens.

Frostbite

This becomes noticeable when the tips of the comb – especially in those breeds with larger headgear – become blackened and drop off due to the tissue dying as a result of freezing temperatures. You can try to pre-empt it by applying petroleum jelly or similar grease to the comb and wattles in the evening when you know that the temperature is due to drop below freezing.

Diarrhoea

Besides a runny bum caused by pathogens, diet excesses can result in similar symptoms (not to be confused with the caecal dropping, which are the smelly, foamy mass expelled roughly every seven to ten droppings) – for example, excess cabbage or even a sudden change to a different diet. See also the section on bacterial infections on pages 139-42.

Below: Brown, foamy and perfectly normal caecal dropping.

Above: The nictitating membrane is the 'third' eyelid.

Eye trauma

This is most often seen when a feather grows down into the eye in heavily crested birds, although other causes include pokes, scratches and insect bites in and around the area. Infections may also cause trauma, for example sinusitis and mycoplasma, where swollen tissue presses on the eyeball. Antibiotic eye drops are helpful but may be difficult to administer, as the bird's third eyelid (nictitating membrane) gets in the way. A wax-based animal eye balm, which can be prescribed by a vet, is a better alternative.

An excess of ammonia in a house that has not been cleared out for a while will also cause eye problems, as well as respiratory difficulties, which in turn can cause the cornea to become opaque. Certain strains of the Marek's virus will also cause 'grey eye' and you should look to see if any other symptoms are present.

Aspergillosis (fungal)

This is a zoonotic but rare disease, known as Farmer's Lung when it occurs in humans. It's difficult to treat in chickens, as by the time symptoms are exhibited it has usually spread throughout the body, causing respiratory distress and infecting the bones and abdomen. Avoid mouldy or decaying plant matter, as aspergillosis is caused by inhaling fungus spores. Healthy birds will cope with a small level of infection, but may die if they become stressed. It's lethal in chicks, which has earned it the alternative name brooder pneumonia. Avoid alternating wet and dry conditions, which will aid spore production and then dispersal. Provide good ventilation and avoid overcrowding.

Favus (fungal)

This is an uncommon fungal infection that affects the skin. Symptoms include a white, powdery coating on the comb that resembles flour (whence the infection's traditional name of white comb). In severe cases this may spread to the neck and resemble a honeycomb, as feathers fall out and the skin takes on a yellow, rotting appearance. A topical fungicide from the vet is effective.

Heart attack

Larger birds and heavy breeds are particularly susceptible to heart attacks, although they're by no means exclusively affected. The first sign may be an intermittent bluish tinge to the comb without any other symptoms, and unfortunately such birds die suddenly. Turkeys are especially susceptible due to high blood pressure, with reports of commercially reared poults being scared to death even by heavy rain and thunderstorms.

Water on the brain

This is a particular problem with, although not limited to, some of the crested breeds, and is often mistaken for Marek's disease. An infection in the enlarged cranial cavity produces fluid that in turn presses on the brain. Symptoms include walking backwards and falling over. The bird may spin around in circles then suddenly recover, much to the misguided amusement of the keeper.

Treatment is simple and effective: isolate the bird and administer a liquid feed (pellets ground up with water) directly into the mouth via a large syringe at regular (two- to four-hour) intervals during the day, with the last feed at night. An antibiotic preparation needs to be prescribed by your vet and injected into the breast muscle along with an anti inflammatory drug (ask your vet to show you how if you're unsure), to reduce the swelling and clear up the infection.

You may have to continue treatment for up to four weeks in mild cases. However, severe cases should be culled if there's no marked improvement after one week or if the bird is in considerable distress.

Above: Dirty drinkers can cause disease.

Below: Water on the brain can be distressing when first witnessed.

Below: Intra muscular injection should be demonstrated by your vet.

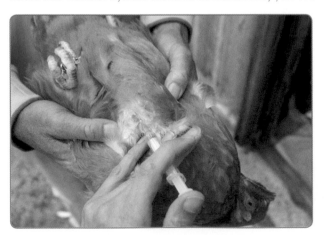

Poisoning (plants, chemicals)

Fortunately most poisonous plants are bitter and when crushed usually emit an odour, and chickens won't attempt to eat them. Watch out, though, for blue-green algae, which, although appears harmless, is quickly fatal, so drinker hygiene is essential.

The following common plants may be a problem where you live:

Below: Laburnum.

Above: Members of the solaniacae family have similar flowers.

Above: Foxglove.

- Many members of the pea family, such as laburnum, sweet peas and vetches.
- Members of the family solaniacae – potato foliage and flowers, tomato foliage and flowers, nightshades and their berries.
- Privet and its berries.
- Yew.
- Rapeseed.
- Digitalis (foxglove).

Chemicals are no less dangerous, as although you may keep them in a shed, vermin and other intruders may damage their containers, thereby allowing your chickens access. Some sources of poison aren't as obvious as you might first think:

- Fungicides such as Cheshunt compound contain copper, which will cause convulsions and in severe cases death.
- Lead from paint and lead shot causes green diarrhoea and weight loss; there may also be some associated head swelling. Similarly, zinc from corroded drinkers and feeders will cause weight loss and weakness in the legs. The recognised treatment is sodium calciumedetate, which should be given by a veterinary surgeon and is effective in most cases.
- Matches contain phosphorus, and ingestion of the heads can cause sudden death or weakness. Fireworks are another source of phosphorus (as well as many other poisonous compounds) and should never be stored; it's better to buy them and use them on the same day and then pick up any debris.
- Nitrates from fertilisers result in increased thirst and a purple comb. Excess can be treated intravenously with methylene blue, again administered by a vet.
- Sodium chloride (common salt) causes kidney failure and convulsions, and is often picked up in the winter from roadside deposits of rock salt. Keep animal salt licks well

out of the way of chickens, and avoid salty scraps such as cured meat and crisps.
- Fungicides that are often found on seed coatings (blue or pink) are easily ingested because, naturally, the seeds are attractive. They cause retarded growth in youngsters and misshaped, infertile eggs in adults. Large doses are fatal.
- Herbicides can cause convulsions and death, so avoid putting chickens out on a lawn that's been treated (a common mistake).
- Malathion, an organophosphorus compound, accumulates in the body, taking many weeks before the bird dies of vomiting, possibly with associated muscle spasms or tremor. Many mite-killing preparations contain this but are now being withdrawn owing to concerns over human health.
- Creosote, once used to preserve poultry houses and kill red mites, but now banned, still occasionally crops up in old sheds and kills young birds. Thoroughly wash any drinkers or feeders that have been cleaned with compounds containing phenol.
- Ammonia from dirty house litter can cause eye problems, and damages the cilia (small hair fibres) in the bird's windpipe, which in turn means that the mucus that would normally trap and move pathogens to the stomach can't flow properly and the bird becomes exposed to infection.
- Carbon monoxide is a by-product of fuel burned in insufficient oxygen, but it's encountered by poultry keepers less often now that brooder rooms use electric heaters rather than gas. It bonds preferentially to the body's red blood cells, limiting the amount of oxygen that can be absorbed. Dead birds will have bright, cherry red blood.
- Slug pellets are probably one of the most dangerous products in the garden shed, as they look and feel like feed pellets and contain metaldehyde, which is lethal to chickens.

Above: Chickens will roost at night to sleep.

Not perching

The natural instinct of a bird is to roost higher than they nest, so if your cosy, darkened nest boxes are higher then this is where your hens will end up, as it's a question of security rather than comfort. Newly acquired birds may not be used to a perch, especially if they've been commercially reared at a hatchery, and you may need to lift them on to the perch at dusk (where they'll stay quite happily overnight) until they get the hang of it. Rescued battery hens usually always need teaching, but they soon get the idea as their natural instincts come to the fore.

Failure to perch may not necessarily be the bird's fault, though, as mites may be the problem – they crawl along the perch to get to their meal, and the chickens soon associate perching with itching. So check the perch sockets and treat any infestation, especially if the birds don't want to go into the house at night.

Below: De-beaking is to be discouraged.

Check, too, that the perch is stable, as nothing will put a bird off quicker than a wobbly perch. Also ensure that it's suitable, not something like an old broom handle (which is too round, and the birds will slip off). The perch needs to be about 5cm (2in) wide when chamfered at the edges (to allow claws to curl round it); an 8cm x 8cm (3in x 3in) length of wood bevelled along two sides does the job nicely.

Stress

In order for you to keep your birds in good health for eggs, table or even manure production, it's essential to minimise stress, which encompasses anything that might reduce their resistance to disease. Many disease-causing microbes are actually benign and only cause illness when the birds' resistance is reduced. Some diseases are stressful in themselves, and while not inherently dangerous they lower the birds' resistance to more serious infections; something like a worm burden, while not necessarily dangerous in itself, can reduce disease resistance and open the way to something far more serious.

A bird's life is full of stressful events, such as hatching, reaching maturity, laying and moulting, and it's not by coincidence that many diseases are specific to a particular age range. Additional factors such as cold, heat or wet, or poor management such as excessive handling or de-beaking, will all have an impact.

Stress management involves providing clean, dry bedding and accommodation, adequate protection from the elements (including freedom from draughts), good food and water supply, and space. Avoidance of crowding is especially important for youngsters, since as they grow they develop immunity through gradual exposure to disease challenges, and overcrowding causes stress which manifests as disease. Avoid, too, moving a flock that's just coming into lay, or during extremes of weather.

Below: Your chicken will appreciate gentle handling.

According to a recent study, gentle handling of birds is more important to their health than a pretty house, and compared to birds that are treated roughly, those treated gently grow to a more uniform size and are less prone to infection – which is more reason, then, not to pull a bird by its legs in the manner favoured by many bad handlers.

An easy way to reduce stress is 'little and often', or preconditioning as it's also termed. For example, if you intend to move a laying flock, do so a few weeks before you need eggs, to allow the birds to acclimatise to their surroundings; or if you have new birds to introduce do it gradually, perhaps with a separate run in the main run, and after a few days put the birds in with the main flock at night. All said and done, if you reduce the stress you improve the health of your birds, and if you want to keep the birds for the garden, then a happy bird is a healthy and long-lived one.

Heat stress

Heavy-feathered breeds such as Cochins or Brahmas, as well as birds in full lay, are quickly affected by this, as are darker birds. Don't 'trap nest' during hot weather (the process whereby a hen is blocked into a nest box to determine which one has produced a specific egg for breeding purposes), as an enclosed bird will quickly succumb.

Having no sweat glands, a chicken's response to heat is to pant and raise its wings to increase the surface area exposed to the air for cooling. An adult bird will start to pant when the temperature reaches 29.5°C (85°F), and chicks when the temperature is 38°C (100°F.) If the temperature reaches 40°C (104°F) panting is insufficient and deaths will occur. You can help towards preventing this by increasing ventilation and supplying fans for indoor birds, or by providing shade for outdoor birds. Heat stress can cause secondary problems such as cannibalism, feather pecking, and toe pecking in chicks.

Below: A hen will pant to cool down.

Cannibalism

This is more frequent in lighter breeds such as Leghorns, less so in the heavier Asiatic breeds. Causes include:

- Heat stress.
- Insufficient darkness in which to lay.
- Bright light.
- Overcrowding.
- Boredom.
- Too little feed/water.
- Too many calories in feed.
- Parasites.
- Injury.

As soon as blood is seen, the other chickens will peck relentlessly until the unfortunate victim is fatally injured. For solutions, see the section on feather pecking at the beginning of this chapter.

Broodiness

While those of you breeding for the table or for replacements will no doubt welcome nature's way of hatching, those who are trying to maximise egg production won't, as while a chicken is sitting she's not laying. Spotting a broody is easy: she looks well but refuses to come out of the nest box, if disturbed she squawks and fluffs up her feathers and either shuffles around or pecks you, and if you pick her up out of the nest box she'll sit indignantly on the floor clucking and will often defecate a large stinking mass, which isn't pleasant.

Broodiness will last approximately three weeks, the time it usually takes to hatch chicks, although it can last for longer.

Below: A determined broody will squawk and fluff out her feathers.

Above: Broody droppings may be loose and foul smelling.

Above: You should be able to recognise if your birds are content.

Above: An anti broody coop is designed to discourage sitting.

During this time it's absolutely critical to move your bird to feed and water. Many methods have been tried to get the hen out of her broody state of mind, so once you've found an effective (and humane) method it's wise to stick to it. One of the best is to move her to a specially made anti-broody coop, which has a wire base to encourage airflow around her nether regions to cool her down (her breast being hot is another sign of broodiness) and a slatted front to make it necessary for her to get up to reach food that's temptingly close.

Ensure that the anti-broody coop is housed in an area not accessible to foxes or rats, as your hen won't be able to escape and the very nature of the exposed house makes it easy to break into. Limited success can also be achieved by putting the culprit in with a cockerel, as often his amorous attentions will do the job – but you need to be on hand to break up any serious fighting.

Simple diagnosis

If you've assembled a fairly effective first aid kit then this will be suitable for most minor complaints such as cuts and sniffles. However, if you suspect something out of the ordinary then you should always consult your vet. Although some veterinarians have only a limited working knowledge of poultry they do have access to pathology labs for accurate diagnosis, which is invaluable if you've a serious outbreak of something nasty running through your flock.

Familiarity and knowledge of your own birds is essential if you want to be able to accurately assess symptoms. For example, your birds may be used to dogs, but your neighbour's may suddenly go off lay when they see your border collie – in other words, both flocks will exhibit entirely different reactions to stress.

Below: A hen will squat automatically.

Above: Having placed your hand on the chicken's back, slide your free hand under the breast.

Handling

This should be done regularly, in order to familiarise yourself with how your bird feels: its weight, shape, and if it's lively or placid.

Never pick up a chicken by its legs or wings, as you can do serious damage and cause undue stress. It's an old practice that was used to collect as many birds in one go as possible, but it needs to be discouraged – although unfortunately it's still being taught on some husbandry courses. The correct and much more humane way is to pick up and support the bird on your arm or hand.

First approach the chicken and put a hand firmly on her back while pressing down lightly; this induces a squat, as the chicken thinks she's being trodden on by a cockerel. Some hybrids are actually bred specifically to squat when you approach them, to make handling that much easier.

Slide the fingers of your free hand under the bird, and with your fingers on either side of her legs and your middle finger between her legs, pick her up.

Her head should point towards your elbow, her backside away from you. This is for good reason, as a nervous bird will defecate instantly.

Small birds will balance nicely in your hand, while larger ones will lie along your forearm.

Above: Pick your bird up.

Below: Grip the legs between your fingers.

TOP TIP

If you have a nasty cock bird, wear suitable protection on your arms and hands. Pick him up and stroke him if he charges at you – since he's expecting retaliation, this will often be enough to shock him into quietness.

Monitoring

By regular monitoring of your birds you'll come to know what's normal and what's abnormal behaviour: a bird lying on its side and thrashing around in circles may be normal behaviour if it's sunny, but abnormal if there's no sun and the bird doesn't stand up when you approach. Keeping daily notes is even better, since when a problem becomes apparent all of those little things you've been noticing will suddenly fall into place.

This monitoring will quickly build up a 'flock history', and by knowing how a fit bird looks, smells and acts you can add in the visible signs, working from top to toe, which will enable you to make a better diagnosis when symptoms are similar, and to differentiate between problems that are merely troublesome and those that are potentially fatal.

Above: A swollen face tells you this bird is not well.

Body part	Preferred appearance
Head and tail	Held high and straight, not drooping or bent over to one side. Fully feathered.
Comb & wattles	Bright, waxy, vibrant colour when in lay or mating, warm, pliable, free from coatings or flaking.
Eyes	Clear, bright, shiny, pupil gets bigger in shade, smaller in light. Eyelids clear from mucus and foam. Third eyelid can't be seen when resting.
Nostrils	Clear, no rasping, free from dust and mucus. Dry.
Crop	Firm when full, disappears when empty.
Breast	Full, plump, free from blisters and growths. The sternum should still be able to be felt, although not prominent, and should be free from dents and kinks.
Abdomen	Firm, but not hard, should not feel spongy like a hot-water bottle.
Posture	Erect, alert, active. The bird should be constantly moving. Inquisitive and wary.
Feathers	Clean and lustrous. Held tight to the body (depending on breed). The bird should preen regularly.
Vent	Clean, moist, no signs of parasites or scabs. It should pulse when inspected.
Droppings	Firm, grey/brown with a white 'cap'. Caecal droppings are foamy, light or dark brown and smelly.

When you're familiar with how your birds should look, consider any exceptions that there may be:

■ From a distance, note how the birds move as a flock. Is any particular bird left behind or being picked on? When the birds stop, how do they look? Can you hear or smell anything unusual when up close?

■ How many birds are affected by a problem? How fast are the symptoms spreading – hours, half-days, days, weeks?

■ Are birds dying? If so, how long from the onset of symptoms? What percentage of ill birds die?

■ Pick up a bird. Is there any abnormal discharge from any area? If so, note the colour, smell and consistency.

■ Are there any signs of wounds, parasites, swelling or blindness? (A bird will blink if you bring a finger slowly towards its eye – but be warned, air movement will make a bird blink too.)

■ Is there a change in the bird's temperature? Insert a thermometer gently into the vent. Normal temperature for a chicken is 40°C (103°F), for a chick 42°C (106.7°F).

Below: A sunbathing chicken may appear to be having a seizure.

Above: Flock history will tell you if foam in the corner of the eye is a result of physical trauma or infection.

The bird in the above photo might have been suffering from ILT, mycoplasma, a simple eye infection or infectious bronchitis amongst many other things, but by knowing the flock history and the onset of symptoms it was possible to assess the likely cause more accurately – which in this case turned out to be a feather in the eye, since there were no other symptoms.

Common symptoms

Many diseases have common symptoms such as ruffled feathers, weight loss and reduced egg production. Some viral diseases, however, have symptoms common to groups, for example coughing, wheezing, rattling and general laboured breathing are all symptoms of respiratory disease, while diarrhoea, off-coloured or bloody droppings, thirst (wet heads and bums may indicate increased water intake) and dehydration indicate intestinal disease. Problems with

balance and walking may be skeletal or muscular, with twitching, trembling and convulsions indicating nervous system diseases.

In addition there will be specific symptoms produced by individual, normally bacterial diseases, such as a hard, swollen abscess on the foot pad, which is specific to bumblefoot.

- Weight loss can be caused by mites, worms, bullying, TB, coccidiosis or any disease affecting the intestines, lack of water or a heavy moult.
- Diarrhoea can occur as a result of salmonella, coccidiosis, E. coli or excess cabbage.
- Failure to produce eggs can be due to IB infection, stress, egg eating, moult, age, day length or obesity.
- Wheezing can be the result of mycoplasmas, IB, aspergillosis, ammonia exposure or ILT.

Notifiable diseases

These are the diseases that, due to their nature and ease of transmission, constitute a threat to human health or economics, and must be reported to DEFRA (the Department for Environment, Food and Rural Affairs).

DEFRA state that: 'Any person having in their possession or under their charge an animal affected or suspected of having one of these diseases must, with all practicable speed, notify that fact to a police constable.'

In practice, if you suspect signs of any of the notifiable diseases in the table below, you must immediately notify a DEFRA Divisional Veterinary Manager (go to http://www. DEFRA.gov.uk/animalhealth/about-us/contact-us/search/index. asp for an interactive map and contact details).

Notifiable disease	Species affected	Last occurred in UK
Avian influenza (Bird 'flu)	Poultry	Present
Newcastle disease	Poultry	2006
Paramyxovirus of pigeons	Pigeons	Present

Below: Eyes should be bright and alert.

Below: Glossy and red, the face of a healthy bird.

Biology and anatomy

Essential data

Temperature For an adult chicken around 40°C (103°F), for a chick around 42°C (106.7°F).

Heartbeat 250–350bpm in adults, 350–450bpm in chicks, depending on breed.

Respiration Cocks 12–20 breaths per minute, hens 20–36 breaths per minute.

Immune system

This is the first barrier that any disease or parasite encounters in the skin, and the largest of an individual animal's organs. Access to the bird's system can only be attained by penetrating this layer or through the mucous membranes, which are well defended by mucus and other fluids. Should any of the natural bacteria that inhabit the skin, or any

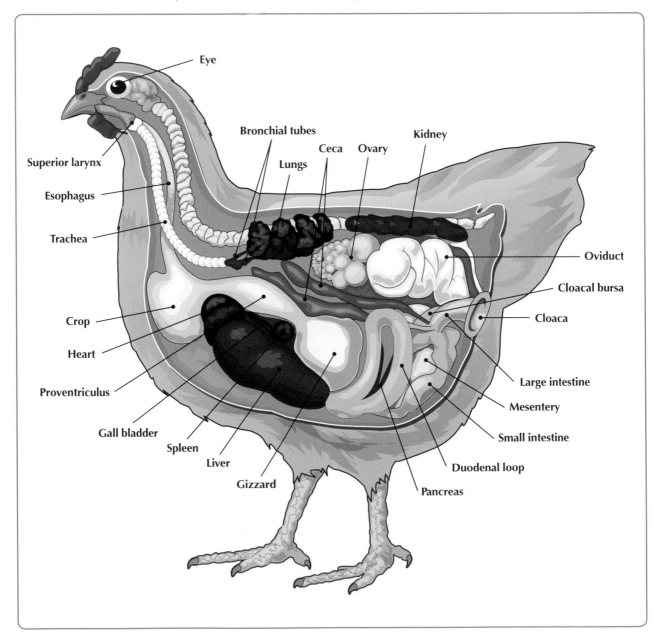

Eye

Bronchial tubes

Lungs

Ceca

Ovary

Kidney

Superior larynx

Esophagus

Trachea

Oviduct

Cloacal bursa

Cloaca

Crop

Heart

Large intestine

Mesentery

Proventriculus

Small intestine

Gall bladder

Spleen

Liver

Gizzard

Duodenal loop

Pancreas

Above: The eye is often the site where bacteria and viruses enter the body.

Above: Healthy birds will ingest stones to aid digestion.

invaders enter the system, then the killer white blood cells that inhabit the lymphatic system take over and engulf the foreign matter (antigens). All cells in the body (ours included) are bathed in 'lymph', which is produced in the bloodstream and circulates through the lymph nodes (or glands), which remove the foreign matter. It's only when these antigens overcome the lymphatic system and multiply in the body of the chicken that an outbreak of disease occurs.

In chickens, immunity and antibody production are controlled by an organ specific to chickens called the cloacal bursa, also known as the *bursa of Fabricius*. This is situated internally, just behind the cloacal opening. Any disease that disables or damages this organ, such as infectious bursal disease or Marek's disease, causes immune-suppression, so that the bird becomes easily infected by other agents. These secondary infections mask the original problem and often prove fatal.

Digestive system

This is literally the chicken's power unit, where food is converted into energy to run the body's systems. The process starts with the bird's beak, which picks up and swallows its food. If you open the beak and look at the roof of the mouth you'll notice that it's quite ridged, which has the effect of holding food in place with the aid of the muscular tongue. Saliva is mixed with the food, which is then swallowed.

The package of food then passes along the oesophagus (food pipe) and into the crop, where mucous glands excrete a lubricant that softens the food and the crop becomes distended. (This is why show birds aren't fed on the morning of a show, to prevent distension and the ruining of the profile.) The food is then passed through the glandular stomach (proventriculus) to the muscular stomach (gizzard) and back again several times. The proventriculus adds gastric juices while the muscular gizzard grinds the food with the aid of swallowed grit. The resulting mush then passes to the duodenum for further digestion and then along the intestines for absorption.

Before reaching the large intestine, a pair of caeca (also termed the blind gut) are encountered, where cellulose is fermented by bacteria for absorption. Contents from the large intestine and the caeca (every seventh to tenth dropping) are voided into the rectum and expelled through the cloaca. A normal dropping is firm and brown with a white tip of urates from the kidneys, and is easily distinguished from the foamy brown caecal dropping.

The liver is situated next to the duodenum along with the gall bladder. As well as vitamin storage and detoxification of the blood the liver provides the fats that are destined to form the egg yolk. If punctured when preparing a bird for the table, the gall bladder will leave a bitter taste in the meat.

Excretory system

A chicken's body is very much geared to preserving water, hence the fact that chickens don't urinate, but rather exude solid white matter along with their digestive droppings. As its spinal nerves pass through the kidneys, any damage to a chicken's kidneys may manifest itself as lameness.

Below: Discoloured urates may indicate an internal problem.

Respiratory system

Although subject to many problems this is probably the easiest of the chicken's systems to treat. The nasal opening functions as a heat exchange system and helps to reduce water loss, aided by the chicken panting and fluttering its throat (a process known as 'gular flutter'). Because this area is reduced by the operculum, a horny flap of skin, it's easily blocked by any swelling.

The sinuses sit in front of the eye just below the skin and easily swell if an infection sets in. This in conjunction with foam in the corner of the eye is an early indication of a respiratory disease.

As the bird breathes in, air passes along the trachea and into caudal air sacs, before being passed into the lungs where its oxygen is taken up in the bronchi (small air passages) and then passed into the cranial air sacs. On breathing out, the process is reversed. Because of the nature of the respiratory system, any infection in the lungs is soon passed deeper into the body, where it becomes more difficult to treat.

> **Fact...**
>
> **A chicken's sense of smell is comparable to our own.**

Above: Caponising (sterilising) a cockerel is now illegal in the UK.

Reproductive system

Birds come into breeding condition when days begin to lengthen after a period of being shorter, a term known as photoperiodic. It therefore follows that if you want your hens to continue laying, you should ensure that they experience some degree of decrease in daylight so that they'll respond correctly to you increasing the light in the hen house by artificial means.

The male

In the cock bird, the testes are located close to the spine between the lungs and the kidneys. During mating the sperm is passed via a groove in the vent into the vent of the hen, where it is then stored in glands in the vagina, remaining viable normally for around seven days, which is the minimum period of time you should wait if you wish to change a cock and be sure that any subsequent progeny are from the replacement.

Caponising (sterilising) a cockerel at eight weeks of age used to be achieved by inserting a capsule of oestrogen behind the comb, which effectively neutered the bird. This is now illegal in the UK, along with castration.

The female

It's the hen that determines the sex of its progeny, having XY chromosomes. Normally only one side of her reproductive system (usually the left) is functional, the right being atrophied. Although the female embryo starts off with both ovaries only the left one develops fully, although occasionally the right will also develop partially, forming a cyst that causes her death by interfering with her other organs.

From start to finish the entire process of laying an egg takes approximately 25 hours, which means that a hen will lay later and later every day until such time as she skips a day – you'll particularly notice this if you have four hens, because you won't get four eggs every day. This is due to ovulation taking place about half an hour after laying, and if it's dark by the time this is due to happen then no egg is released from the ovary until the following day.

Light breeds come into lay at about 18–20 weeks, and heavy breeds normally at around 26–28 weeks, depending on the time of year. Point of lay is traditionally 18 weeks, and this is an age not a stage, so you'll need to wait an appropriate number of weeks extra before laying will commence. It was at this age that birds used to be moved from their growing pens to their laying houses.

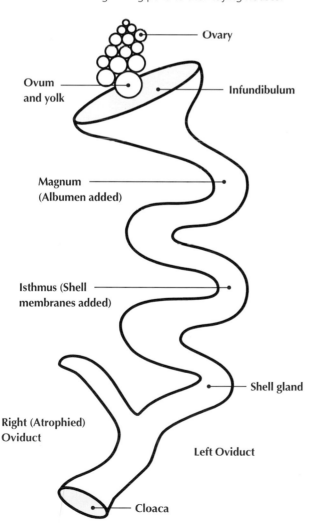

- **Ovary**
- **Ovum and yolk**
- **Infundibulum**
- **Magnum (Albumen added)**
- **Isthmus (Shell membranes added)**
- **Shell gland**
- **Right (Atrophied) Oviduct**
- **Left Oviduct**
- **Cloaca**

Above: Comb is pale pink, about 4 weeks off lay.

Above: Comb is dark pink and larger, about 1-2 weeks off laying.

Above: Comb is red, larger with definite serrations, laying is imminent.

Above: The pineal gland detects light information.

Endocrine system

Of the chicken's chemical system, it's the pineal gland that's of the most interest, as it's the one which responds to daylight and is responsible for the breeding hormones and, therefore, laying. Optimum day length is 14 hours, and any lighting increase should be started from November, when the chicken has had a chance to experience reduced day length (see above). One 40W bulb is sufficient for 2m² (21sq ft) as long as you can read by it, and you should increase the light gradually over a two-week period, adding the extra in the morning in order to allow the birds to perch naturally in the evening.

Skeletal system

As with every other animal, this is the system that supports the organs, holding them in place and allowing mobility. In addition, the skeleton of a chicken helps with respiration via hollow pneumatic bones, and with calcium delivery via special medullary bones.

When perching, the bird crouches on a suitable surface and its legs bend beneath it, which tenses the tendons so that the toes automatically bend round the perch.

The main problems associated with the skeleton are usually either as a result of calcium deficiency due to deficiency in the diet, when disease interrupts uptake and absorption, or as a result of physical injury, possibly by poor handling. Inflammation occurs occasionally, when a bird will develop a swollen joint (normally the

hock) that feels hot to the touch. This is caused when the synovial membranes that normally protect the joint become inflamed as a result of disease or injury and produce excess fluid to cushion the joint. The resulting excess fluid is what causes the symptoms, and is best treated with an antibiotic.

Breast blister is caused by a similar process and resembles a water-filled spot on the breastbone. Left untreated, a permanent dent will occur in the keel. This condition is seen in birds when continued pressure is exerted in the breast, possibly due to weak legs that can't support the bird when perching, or if there are insufficient feathers present.

Below: Skeleton.

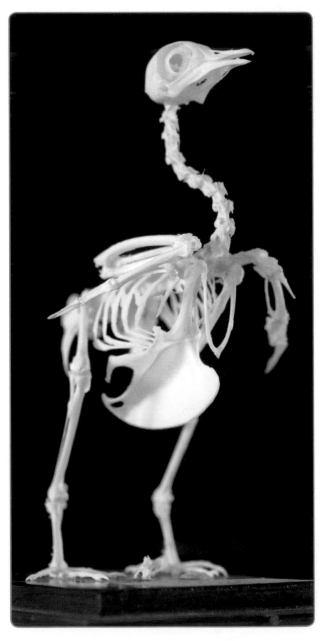

Above: A suitable perch will prevent breast problems.

Nervous system

The central nervous system controls voluntary movements (eg eating) and the peripheral nervous system controls involuntary movements (eg breathing). The beak is a sensory organ used to feel food, and as such contains a large number of nerve endings, so the practice of de-beaking, where the beak is cut to prevent pecking in battery hens, is to be discouraged, as it's very painful to the bird. Similarly, the tail – which is the bird's steering mechanism – also contains many nerve endings, which are quickly affected by illness; drooping of the tail is therefore often the first indication of a problem.

Medical issues such as Marek's disease, poisoning or genetic factors can all have an effect on the nervous system. Symptoms typically include lack of co-ordination, trembling, convulsions or paralysis.

Above: The tail contains many nerve endings.

Circulatory (cardiovascular) system

This is the blood and fluid system of a chicken. It is controlled by the heart, which is proportionally larger than a mammalian heart due to the demands of flying, and a rapid heartbeat and higher blood pressure result in much better oxygen transportation throughout the body. In mammals the lungs surround the heart, in chickens the liver does.

The carotid arteries are well protected by the neck vertebrae and continue to function even if the throat is cut, which therefore means that this is a very inhumane method of dispatch – the more acceptable method is dislocation of the neck, which severs the arteries. As was mentioned earlier, the beak, claws and spurs all contain blood vessels that are easily damaged if cut incorrectly.

Below: The outward sign of a healthy circulatory system – a bright red face.

FURTHER INFORMATION

Cooking guidelines

When cooking or preparing any type of poultry or eggs, it's always advisable to work cleanly and avoid contaminating other work surfaces and utensils. Having an antibacterial kitchen spray handy is always an ideal solution.

By following these guidelines, you'll minimise the risk of contracting food poisoning.

Handling

- In order to avoid contamination of the meat by feathers or droppings, when dressing the carcass rinse it thoroughly in clean, cold water.
- Thaw frozen poultry thoroughly in the bottom of the refrigerator overnight, never on a work surface.
- Don't defrost on a shelf over other foods, especially cooked meats.
- Wash your hands and any utensils (including the chopping board) with an antibacterial washing agent. Ordinary domestic bleach is a good substitute, diluted in accordance with the instructions on the container.

- Don't let cooked meat come into contact with raw meat or the utensils and surfaces used to prepare raw meat.
- Collect eggs regularly and refrigerate them if they're not going to be used immediately.
- Discard any cracked or soiled eggs.
- Wash hands and utensils after preparing eggs.
- Avoid serving foods containing raw egg to the very old, young or infirm, and especially to those with a compromised immune system.

Cooking

- Heat meat to 85°C (185°F).
- Reheat leftovers to 74° (165°F).
- Cook casseroles and other poultry-based meals to 71°C (160°F).
- Cook all meat until no red shows at the joints and when a skewer is inserted the juices run clear.
- Heat eggs and dishes containing eggs to 71°C (160°C).

Below: Use only clean, unbroken eggs.

Useful egg recipes

Boiled eggs

In order to boil an egg to perfection, you must firstly apply a few useful tips to ensure that you get the most out of the process:

- Use a small saucepan to stop the eggs rolling around and cracking.
- Place the eggs in cold water, covering them by about 25mm (1in).
- Bring to the boil then immediately reduce the heat to a simmer. Do not continue to boil at high heat, as this turns the yolk to rubber and a grey/black colour.

TIME FOR

3 minutes – very soft-boiled.
4 minutes – creamy yolk and just set white.
5 minutes – solid yolk with a soft middle, set white.
7 minutes – hard-boiled.

To avoid your eggs cracking when they're boiled, bring refrigerated eggs up to room temperature first, as the sudden increase in temperature will otherwise crack them. Also, prick the rounded end where the air sac is situated to allow steam to escape, and use a timer.

If you're cooking hard-boiled eggs then use ones that are over five days old, otherwise they're difficult to peel, and run them under cold water for approximately two minutes at the end of the cooking time. This ensures that they stop cooking and prevents them getting a black ring around the yolk and an increased smell of sulphur.

Below: The perfect boiled egg will have a creamy middle and no sign of a black ring.

Calorific values

An egg is the ideal packed lunch and is power packed with nutrients:

Nutrient	Units	per 100g	Large (63-73g)	Medium (53-63g)	Small (less than 53g)
Water	g	75.84	37.92	33.37	28.82
Energy	kcal	143	72	63	54
Protein	g	12.58	6.29	5.54	4.78
Fat	g	9.94	4.97	4.37	3.78
Ash	g	0.86	0.43	0.38	0.33
Carbohydrate	g	0.77	0.39	0.34	0.29
Fibre	g	0.0	0.0	0.0	0.0
Sugars	g	0.77	0.39	0.34	0.29
Minerals					
Calcium (Ca)	mg	53	26	23	20
Iron (Fe)	mg	1.83	0.92	0.81	0.70
Magnesium (Mg)	mg	12	6	5	5
Phosphorus (P)	mg	191	96	84	73
Potassium (K)	mg	134	67	59	51
Sodium (Na)	mg	140	70	62	53
Zinc (Zn)	mg	1.11	0.56	0.49	0.42
Copper (Cu)	mg	0.102	0.051	0.045	0.039
Manganese (Mn)	mg	0.038	0.019	0.017	0.014
Fluoride (F)	mcg	1.1	0.6	0.5	0.4
Selenium (Se)	mcg	31.7	15.8	13.9	12.0
Vitamins					
Thiamine	mg	0.069	0.035	0.030	0.026
Riboflavin	mg	0.478	0.239	0.210	0.182
Niacin	mg	0.070	0.035	0.031	0.027
Pantothenic acid	mg	1.438	0.719	0.633	0.546
Vitamin B-6	mg	0.143	0.071	0.063	0.054
Choline, total	mg	251.1	125.5	110.5	95.4
Vitamin B-12	mcg	1.29	0.65	0.57	0.49
Vitamin A	mcg	139	70	61	53
Vitamin E	mg	0.97	0.48	0.43	0.37
Vitamin D	IU	35	18	15	13
Vitamin K (phylloquinone)	mcg	0.3	0.1	0.1	0.1
Cholesterol	mg	423	212	186	161

(Values are averaged based on a number of samples)

Scrambled eggs

To make perfect scrambled eggs, you need to be vigilant over the pan and stir constantly to prevent sticking and burning.

- Crack two eggs per person into a bowl and whisk together with a little salt and pepper.
- Place a heavy-based saucepan over a medium heat and add a knob of butter and a teaspoon of olive oil (to stop the butter burning). Swirl the butter mixture around to coat the sides of the pan to a depth of about 25mm (1in), and then pour the egg in.
- Stir constantly in a figure of eight motion using a wooden spoon until around three-quarters of the egg is set, ensuring that you get into the corners of the pan and cover all of the bottom to avoid hot spots.
- Add another knob of butter and remove from the heat, then continue stirring with a fork until the mixture is set (but not rubbery). It may take you a couple of attempts to master when to remove the pan from the heat.
- Serve immediately.

Below: Keep the eggs moving while cooking.

Poached eggs

Forget swirling water and adding vinegar then dropping the eggs into the vortex – it's a lot easier than that; and throw out the egg poacher and rings that fit in the frying pan, as once you've tried this method you'll never go back to them. Always use a timer, and eggs that are as fresh as possible.

- Fill a deep frying pan with boiled water from the kettle to a depth of approximately 25mm (1in). Heat gently until small bubbles start to rise.
- Crack the eggs close to the surface and slide them into the water and simmer for one minute only.

- Remove the pan from the heat and let it sit, uncovered, for ten minutes.
- Remove the eggs individually from the pan on a slotted spoon, and allow them to rest on a wad of kitchen paper for a few moments to absorb excess water. Serve immediately.

Below: Kitchen towel will absorb the water.

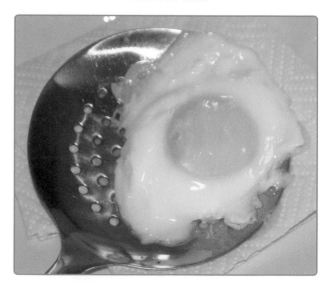

Fried eggs

As with boiled eggs, everyone has their own preference for these, but the following tips will make the whole process easier.

- Use fresh eggs, as they hold together better and the white doesn't run everywhere.
- Use hot fat, and once your eggs are in it lower to a medium heat and leave for about 30 seconds.
- Cook the yolk by using a spoon to baste it with the hot fat for approximately one minute.
- Remove with a slice and drain on a wad of kitchen paper.

Below: Baste the eggs with a spoon to 'blind' the yolk

Pickled eggs

An excellent way to use up excess eggs. Simply hard-boil enough eggs to fit into a sterilised jar, peel them, pack them in, pour in enough malt or white vinegar to cover them (depending on your taste) and seal the jar. It's as simple as that! Try experimenting with added spices and garlic, and with flavoured vinegars.

175

Dispatching

Unfortunately, at some time you may be faced with killing one of your flock, perhaps as a result of illness; and if you're rearing table birds or breeding for exhibition you'll need to dispatch birds as necessary.

If you're not going to eat the bird you'll need to dispose of the body, and you'll need to check with your local council what regulations they have in place for the disposal of stock.

If the bird is destined for the table – and with all the effort that's gone into keeping them you should take advantage of your excess stock – then make sure that you've penned it the night before and given it only water to drink. This is to make sure that its crop is empty before killing and dressing it, since a full crop may fester and spoil the meat during the hanging process. It's also worth giving the bird a dusting with louse powder to make sure that when it's killed the lice don't see you as an alternative home; there's nothing worse than trying to pluck a louse-ridden bird.

1 Remove the bird to a quiet area, out of sight of other chickens as they will become distressed and in turn the chicken in your hands may become agitated and difficult to hold. Hold the bird upside down, with the feet held firmly at waist height by one hand. Grasp the wing feathers (the primaries) with the opposite hand to stop any flapping, and move them to the hand gripping the legs.

2 Hold the bird across your body to the thigh and with your free hand, grasp the chicken's head with the beak and comb against your palm, and the knuckles of your index and middle fingers either side of its neck. Then, while stretching the neck sharply downwards, flick your wrist downwards and you will pull the head upwards and away from the neck cleanly. There will be an audible click which can be sickening to the novice and you should be aware of this. A tip here is to practice with a towel if you are unable to get a real bird to practice on.

Methods of dispatching

Neck pulling

Dislocation of the neck is the accepted and humane method. It is also the quickest method, and relatively clean, with the considerable added advantage that in the case of birds used to being handled, they really don't know what's happening. If possible – if you're doing it for the first time – you should either pratise on a dead bird or have a person experienced in dispatching show you how it's done, in order that you can apply sufficient pressure to dislocate the vertebrae but not so much that the bird is decapitated.

Broom dispatching

This method is suitable for large birds such as ducks and geese or if you have little upper body strength.

- Lay the bird's head on the floor, chin downwards, and place a broom handle or a similar length and thickness of wood over it.
- Hold the bird's feet securely with your hands around the ankles, and put your feet on to the broom either side of the head. Do not at this stage exert any pressure, just position yourself.
- Next, while tilting forwards to put pressure on the broom and secure the head (but not so much that you squash the neck), quickly pull upwards with the bird's feet to break its neck.

3 Once the neck has dislocated, it will create a cavity into which blood can drain when it is hung. This can be made larger by bending the head into the neck.

Dispatching tools

If you prefer, a number of tools are available designed to make the job easier, such as Semark pliers for chicks and small bantams, and wall-mounted dispatchers for larger birds.

Semark pliers are placed behind the head at the top of the neck, and then closed like pliers, smartly, firmly and with no fuss. You can hold the bird calmly in your lap while you do this, but you should use a towel to cover your legs in case the delicate skin tears in the process.

Wall-mounted dispatch equipment is similar to Semark pliers but on a larger and more robust scale. You hold the bird and place its head in the groove. Then, using the lever, press down behind its ears.

If you're preparing to dress the bird for eating, then when done properly the above methods enable the blood to drain into the gaps between the vertebrae and remain inside the skin until the head is cut off during dressing.

Above: Semark Pliers.

Plucking

Dry plucking

It's best to pluck a bird immediately, beginning at the extremities, which cool first. Otherwise the feathers harden in the skin, making the process more difficult – which, incidentally, is why warm water or wet plucking works, as it loosens the feathers.

Like any other process, you'll find a method that you're happiest with, and trial and error will make you familiar with the anatomy of the bird and aware of problem areas. If you do tear the skin don't panic, as you can always put a bit of bacon over that area when roasting, to prevent burning the flesh.

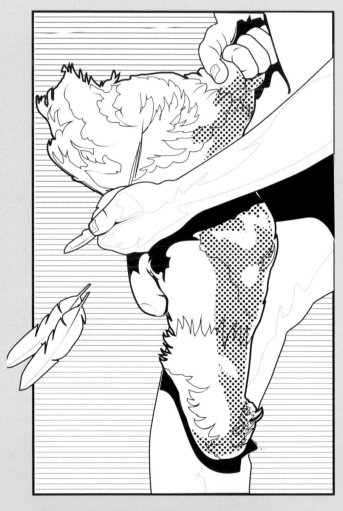

1 Hang the bird so that it is comfortable for you to work on, then after you have suspended it, grasp its legs so that it is secure as you work and does not swing. Begin with the tail feathers and flight feathers on the wings, which can be pulled out with a smart jerk of the hand (you may find gloves help here). If you're strong you can more or less get the whole lot out in one go, but most people will find that the job needs to be approached one feather at a time.

2 Release the bird so that it hangs free and leaves both of your hands free to pluck the rest. Be aware that along the breast and thighs there is a line of slightly longer feathers which are embedded in fat, and these should be plucked next and just a few at a time to prevent tearing the delicate skin. Next move on to the back, which is far more resilient, then the wings, and then the legs and neck.

Skinning

If you only intend to cut out breast meat or are happy with the skin removed from the bird, then plucking is unnecessary. Also, the legs and wings can be removed fully feathered if you don't intend using them.

Skinning is a much simpler option than plucking. Take a sharp knife and simply make a nick under the throat and along the breast and peel back the skin from the entire bird. For just the breast, the meat can be cut away from the bone, rinsed, and either used immediately or frozen.

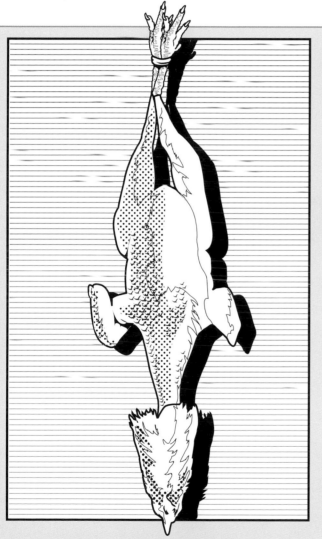

3 Pull the feathers in the direction they're growing, since if you pull the feathers backwards you may again tear the flesh. Finish the plucking process with the breast and then any other feathered areas, leaving the neck feathered about halfway up, as you'll be removing this part later on.

Scalding or dipping

A bird may be scalded by dipping the carcass, head first, into water at about 53°C (127.5°F) for 60–120 seconds, the purpose being to loosen each feather shaft. Hotter water works faster but isn't recommended, as you'll probably have to dip the bird in cold water straight afterwards to prevent partial cooking and the associated problems of food poisoning.

After dipping, re-hang the bird and pluck as above, taking care not to grasp too many feathers at once as the skin will now be more delicate and tear much more easily.

Dressing

After dispatching and plucking, the bird will still be undergoing rigor mortis, where the muscles stiffen, and blood will be draining into the cavity created by dislocation of the neck. You should not attempt to eat the bird on the same day, and dressing is best carried out after letting the bird hang in a refrigerator for four to five days to soften the flesh. Before you hang it in a fridge, make sure you squeeze out any fluids or faeces from the vent, to avoid the area going green.

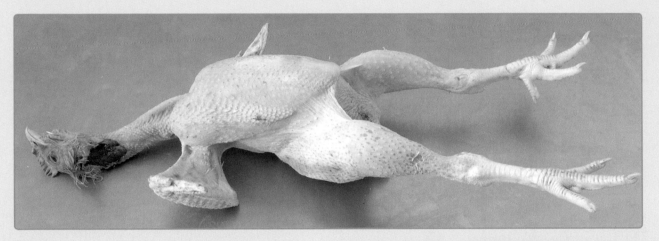

1 There are a variety of ways of removing the legs. The easiest is to use a sharp pair of secateurs and not worry about removing the tendons. Alternatively, use a sharp knife to cut around the point where the scaly leg skin changes into fleshy body skin. Then, having flexed the leg fully forwards and backwards to loosen the tendons, place the leg on a chopping board and apply a sharp downwards force to break the bone. The tendons, which won't have been cut, can then be drawn out. Alternatively, having flexed the leg forwards you can cut through the tendons at the back, then push the leg backwards hard, cut the front tendons, and break the leg away at the joint.

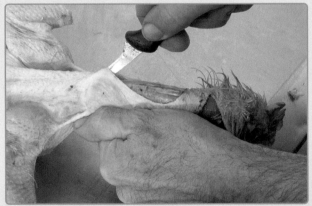

2 Next, make a slit along the skin on the neck.

3 Pull the neck away from the skin, and wind and food pipes.

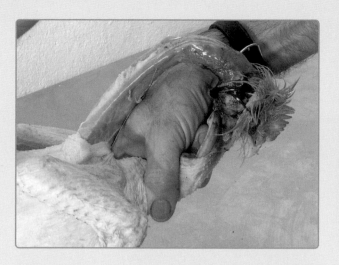

4 Insert fingers into the neck cavity and, using a circular motion, detach the organs from the connective tissue.

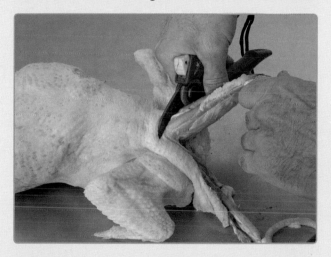

5 Cut the neck away from the body.

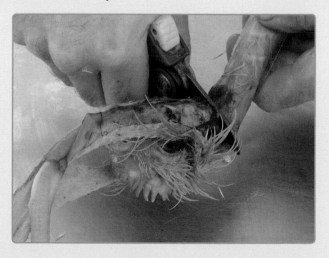

6 Then remove the other end of the neck from the head and keep as part of the giblets if required.

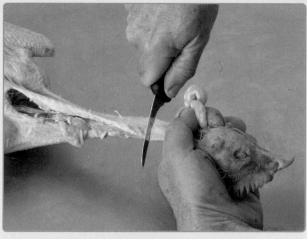

7 Pull out the wind and food pipe as much as possible (still attached to head), and cut away.

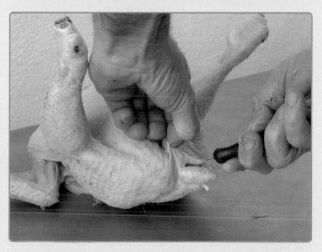

8 Next, put the bird on its back and using either a sharp knife or scissors, pierce the skin carefully just beside the vent and cut around it, making sure that you don't puncture the underlying intestines or rectum.

9 Next cut a straight line from the bottom of this circular cut to the beginning of the bone on the underside, to allow you to get your hand inside the body cavity.

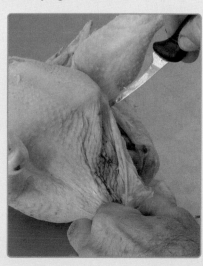

Giblets

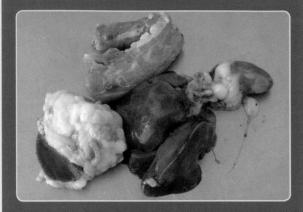

If you want, the gizzard, liver and heart can be cleaned and used with the neck to make a delicious stock for gravy (or chicken livers for a variety of recipes).

A. Split the gizzard down to the white layer and then open up to remove the sack of grit.

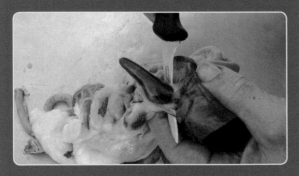

B. Remember also to remove the dark bile sac from the liver at the point where they join – without splitting it – as it will taint the meat.

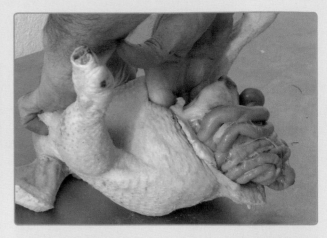

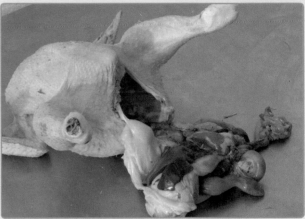

10 Put your hand into the bird and loosen its innards by following a circular path around the body cavity with your fingers. Then you just need to grasp everything with your hand and pull it out of the bird; and having loosened the windpipe from the other end earlier, everything should follow.

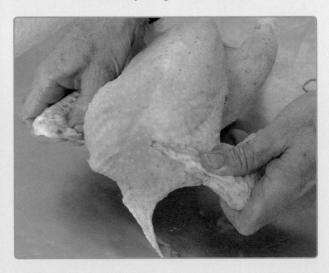

11 Tuck the wings under the bird.

13 Push the legs back into the sides of the body.

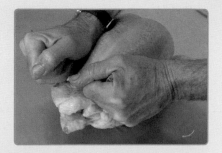

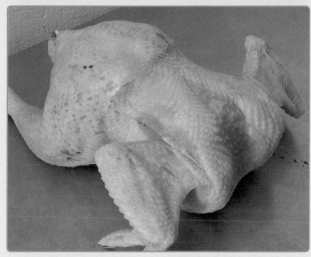

12 Tidy up the neck end and put the loose skin back over the body.

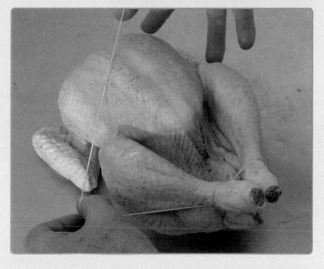

14 Use an elastic band to wrap around the parson's nose and legs then twist over to then wrap around the body. Finally, freeze or roast!

Photographs courtesy of Cracknell's free range poultry farm, Somerset

Contact details

Allen & Page
(Feed supplies)
Norfolk Mill
Shipdham
Thetford
Norfolk IP25 7SD
Website www.allenandpage.com
Email helpline@allenandpage.co.uk
Tel 01362 822900

Ascott Smallholding Supplies
(General equipment)
Units 9/10
The Old Creamery
Four Crosses
Llanymynech
Powys SY22 6LP
Website www.ascott.biz
Tel 0845 130 6285

The British Hen Welfare Trust
(Ex-battery hens)
North Parks
Chumleigh
Devon EX18 7EJ
Website www.bhwt.org.uk
Email info@bhwt.org.uk
Tel 01769 580310

Brinsea
(Incubators)
Station Road
Sandford
North Somerset BS25 5RA
Website www.brinsea.co.uk
Email sales@brinsea.co.uk
Tel 0845 226 0120

Clydesdale Timber Products
(Poultry housing)
Shedyard Farm
Laneside Road
New Mills
High Peak
Derbyshire SK22 4QN
Website www.clydesdale-timber.co.uk
Email info@clydesdale-timber.co.uk
Tel 01663 746784

Cyril Bason
(Day-old chicks and POL pullets,
turkeys and ducklings)
Bank House
Corvedale Road
Craven Arms
Shropshire SY7 9NG
Website www.cyril-bason.co.uk
Email cyrilbason@cyril-bason.co.uk
Tel 01588 673204

The Domestic Fowl Trust
(Poultry, housing and equipment)
Station Road
Honeybourne
Evesham
Worcestershire WR11 7QZ
Website www.domesticfowltrust.co.uk
Email clive@www.domesticfowltrust.co.uk
Tel 01386 833083

Electric Fencing Direct
(Electric fencing)
81 Dunmore Street
Balfron
Glasgow
Lanarkshire G63 0PZ
Website www.electricfencing.co.uk
Email info@electricfencing.co.uk
Tel 01360 440611

Flyte So Fancy
(General equipment including
pop-hole openers)
The Cottage
Pulham
Dorchester
Dorset DT2 7DX
Website www.flytesofancy.co.uk
Tel 01300 345229

Forsham Cottage Arks
(Poultry housing)
Goreside Farm
Ashford
Kent TN26 1JU
Website www.forshamcottagearks.com
Email office@forshamcottagearks.com
Tel 01233 820229

Gardencraft
(Poultry housing)
Tremadog
Porthmadog
Gwynedd LL49 9RD
Website www.gcraft.co.uk
Email sales@gcraft.co.uk
Tel 01766 513036

Interhatch
(Incubators and medical equipment)
Whittington Way
Whittington Moor
Chesterfield
Derbyshire S41 9AG
Website www.interhatch.com
Tel 01246 264620

Littleacre
(Poultry housing)
Botley House
School Lane
Hints
Tamworth
Staffordshire B78 3DW
Website www.littleacre-direct.co.uk
Tel 0121 351 4748

Manor Farm Poultry
(POL pullets)
E.B. & J.H. Franks
Manor Farm
Leasingham
Sleaford
Lincolnshire NG34 8JN
Website www.manorfarmpoultry.co.uk
Email info@manorfarmpoultry.co.uk
Tel 01529 302671

Marriage's
(Feed supplies)
Chelmer Mills
New Street
Chelmsford
Essex CM1 1PN
Website www.marriagefeeds.co.uk
Email sales@marriagefeeds.co.uk
Tel 01245 354455

Meadowsweet Poultry
(POL pullets)
30 Hill Meadows
Durham DH1 2PE
Website www.meadowsweetpoultry.co.uk
Tel 0191 384 2259

Norfolk Poultry Supplies
(General poultry supplies)
37 Langley Green
Langley
Norfolk NR14 6DG
Website www.norfolkpoultrysupplies.co.uk
Email enquiries@norfolkpoultrysupplies.co.uk
Tel 01508 480497

Omlet
(Poultry housing)
Tuthill Park
Wardington
Oxfordshire OX17 1RR
Website www.omlet.co.uk
Tel 0845 450 2056

Parkland Products
(Automatic feeders)
Owley Farm
Acton Lane
Wittersham
Tenterden
Kent TN30 7HL
Website www.parklandproducts.co.uk
Email info@parklandproducts.co.uk
Tel: 01797 270399

Piggotts Poultry Breeders
(POL pullets and day-olds)
16 Wivelsfield
Eaton Bray
Dunstable
Bedfordshire LU6 2JQ
Tel 01525 220944

The Poultry Club of Great Britain
Keeper's Cottage
40 Benvarden Road
Ballymoney
Co Antrim BT53 6NN
Website www.poultryclub.org
Email info@poultryclub.org
Tel 02820 741056

The Rare Breeds Survival Trust
(Conservation)
Stoneleigh Park
Nr Kenilworth
Warwickshire CV8 2LG
Website www.rbst.org.uk
Tel 024 7669 6551

Regency Poultry
(General equipment)
Merrydale Farm
Enderby Road
Whetstone
Leicester LE8 6JL
Website www.regencypoultry.com
Email info@regencypoultry.com
Tel 0116 286 6160

RENCO
(Electric fencing)
Wharf Road
Newton Abbot
Devon TQ12 2DA
Website www.renco-netting.co.uk
Email sales@renco-netting.co.uk
Tel 01626 331188

Rentokil
(Pest control)
Consult website for local contact details
Website www.rentokil.co.uk

Rooster Booster
(Poultry-house lighting)
Barrow Lane Products
Cherry Tree Farm
Charlton Musgrove
Wincanton
Somerset BA9 8HW
Website www.roosterbooster.co.uk
Email info@roosterbooster.co.uk
Tel 07762 298373

Smallholder Supplies
(General equipment)
The Old Post Office
6 Main Street
Branston
Nr Grantham
Lincolnshire NG32 1RU
Website www.smallholdersupplies.co.uk
Email info@smallholdersupplies.co.uk
Tel 01476 870070

Smiths Sectional Buildings
(Poultry housing)
Unit 1
Manor House Farm
Park Lane
Ashley
Market Drayton
Shropshire TF9 4EH
Website www.smithssectionalbuildings.co.uk
Email info@smithssectionalbuildings.co.uk
Tel 01630 673747

Solway Feeders
(General feed equipment)
Main Street
Dundrennan
Kirkcudbright
Dumfries & Galloway DG6 4HQ
Website www.solwayfeeders.com
Email mail@solwayfeeders.com
Tel 01557 500253

SPR Poultry Centre
(Poultry supplies)
Greenfields Farm
Fontwell Avenue
Eastergate
Chichester
West Sussex PO20 3RU
Website www.sprcentre.co.uk
Tel 01243 542815

S&T Poultry
(Chickens, ducks and guineafowl)
39 Windsor Drive
Wisbech
Cambridgeshire PE13 3HJ
Website www.sandtpoultry.co.uk
Email sandtpoultry@btinternet.com
Tel 01945 585618

Verm-X
(Parasite control)
Paddocks Farm Partnership Ltd
Huish Champflower
Somerset TA4 2HQ
Website www.verm-x.com
Email info@verm-x.com
Tel 0870 850 2313

The Wernlas Collection
(Rare breed chickens)
Green Lane
Onibury
Craven Arms
Shropshire SY7 9BL
Website www.wernlas.com
Tel 01584 856318

Woodside Farm and Leisure Park
(Rare breeds, hybrids and general equipment)
Woodside Road
Slip End Village
Luton
Bedfordshire LU1 4DG
Website www.woodsidefarm.co.uk
Email enquiries@woodsidefarm.co.uk
Tel 01582 841044

The breed clubs

The Ancona Club
Phil Smedley
Leckby House
Flaxton
York
North Yorkshire YO60 7QZ
Email phil.boy@virgin.net
Tel 01904 468387

The Australorp Club
Ian and Louise Simpson
Chestnut Farm
Normanton
Southwell
Nottinghamshire NG25 0PR
Website www.poultryclub.org/australorpclubgb
Email ian@australorpclubgb.wanadoo.co.uk
Tel 01636 814958

The Brahma Club of Great Britain
Mrs Sue Black
7 Pleasant Road
Penllergaer
Swansea SA4 9WH
Email brahmaclubgb@yahoo.co.uk
Tel 01792 898310

The British Araucana Club
Mrs Ky Thurland
Tan-y-Rhos
Babell
Flintshire CH8 8PY
Website www.araucana.org.uk
Email araucanasec@araucana.org.uk
Tel 01352 720043

The British Asian Hardfeather Club
Mrs Julia Keeling
Ballashee
Staarvey Road
German
Isle of Man IM5 2AJ
Website www.geocities.com/Tokyo/
Shrine/2425/hardfeather
Email shamolady@manx.net
Tel 01624 801825

The British Barnevelder Club
Mr G. Broadhurst
Little Acres
Chapel Acres
Tern Hill
Market Drayton
Shropshire TF9 3PY
Tel 01630 638630

The British Belgian Bantam Club
Club Secretary
Parc y Rhos
9 Trosserch Road
Llangennech
Llanelli
Carmarthenshire SA14 8AQ
Website www.belgians.jatman.co.uk
Email belgians1@hotmail.com
Tel 07855 074781 and 07891 248938

The British Call Duck Club
Mrs Jen Maskell
Maes y Coed
Llanarth
Ceredigion SA47 0RG
Website www.britishcallduckclub.org.uk
Tel 01545 580425

The British Faverolles Society
Mrs Sue Bruton
Park House
Codsall Wood
Staffordshire WV8 1QR
Website www.faverolles.co.uk
Email park_house@live.com
Tel 01902 843055

The British Waterfowl Association
Mrs Sue Schubert
PO Box 163
Oxted
Surrey RH8 0WP
Website www.waterfowl.org.uk
Email info@waterfowl.org.uk
Tel 01892 740212

The Buff Orpington Club
Phil Smedley
Leckby House
Flaxton
York
North Yorkshire YO60 7QZ
Email phil.boy@virgin.net
Tel 01904 468387

The Call Duck Association
Graham Barnard
Ty Cwmdar
Cwrt-y-Cadno
Llanwrda
Carmarthenshire SA19 8YH
Website www.callducks.net
Tel 01558 650532

The Cochin Club
Andy Marshall
Bradstones
Hewshott Lane
Liphook
Hampshire GU30 7SU
Website www.cochinclub.co.uk
Email administrator@cochinclub.co.uk
Tel 01428 723030

The Croad Langshan Club
Ms L. Heigl
Stillwaters
Thursley Road
Churt
Farnham
Surrey GU10 2LQ
Website www.croadlangshan.org.uk
Email lynh@croadlangshan.org.uk
Tel 01428 602992

The Derbyshire Redcap Club
Mrs J. Louise Woodroffe
1 Alsop Moor Cottages
Alsop-en-le-Dale
Ashbourne
Derbyshire DE6 1QS
Website www.derbyshireredcapclub.org.uk
Email thewoodroffesr@btinternet.com
Tel 01335 310305

The Domestic Waterfowl Club of Great Britain
Michael and Sylvia Hatcher
Limetree Cottage
Brightwalton
Newbury
Berkshire RG20 7BZ
Website www.domestic-waterfowl.co.uk
Tel 01488 638014 (evenings)

The Dorking Club
Mrs Victoria Roberts
Heather Bank
Hillings Lane
Menston
Nr Ilkley
West Yorkshire LS29 6AU
Website www.poultryclub.org/dorkingclub
Email victoriaroberts06@tiscali.co.uk

The Dutch Bantam Club
Mrs C. Compton
Devonia
Northbrook
Micheldever
Winchester
Hampshire SO21 3AH
Tel 01962 774476

The Frizzle Society of Great Britain
Mrs Charlotte Shepherd
26 Eton Road
Oxbridge
Stockton-on-Tees
Cleveland TS18 4DL
Website www.
thefrizzlesocietyofgreatbritain.co.uk
Email tees.frizzles@hotmail.co.uk
Tel 01642 601247

The Goose Club
Mrs D. Moss
Llwyn Coed
Gelli
Clynderwen
Pembrokeshire SA66 7HW
Website www.gooseclub.org.uk
Email contact@gooseclub.org.uk

The Hamburgh Club of Great Britain
Mr P. Harrison
60 Dean Head Summit
Littleborough
Lancashire OL15 9LZ
Tel 01706 377653

The Indian Game Club
John Cook
15 Campton Road
Gravenhurst
Bedfordshire MK45 4JB
Email john@beattietransport.co.uk
Tel 01462 711617

The Indian Runner Duck Association
Christine Ashton
Red House
Hope
Welshpool
Powys SY21 8JD
Website www.runnerduck.net
Tel 01938 554011

The Japanese Bantam Club
Terry and Lisa Crook
Rivermead
Costessy Lane
Drayton
Norwich
Norfolk NR8 6HD
Website web.ukonline.co.uk/japclub
Email japclub@ukonline.co.uk
Tel 01603 868373

The Leghorn Club
Richard Grice
26 Front Street
Staindrop
Nr Darlington
County Durham DL2 3NH
Website www.theleghornclub.com
Tel 01833 660260

The Lincolnshire Buff Poultry Society
Miss Lucy Hampstead
Pine Cottage
4 Station Row
New Bolingbroke
Boston
Lincolnshire PE22 7LB
Website www.lincolnshirebuff.co.uk
Tel 07789 906898

The Marans Club
Mr A. Heeks
44 Poplar Drive
Alsager
Staffordshire ST7 2RW
Tel 01270 882189

The Minorca Club
Rob Walker
Pensons Farmhouse
Stoke Bliss
Tenbury Wells
Worcestershire WR15 8RT
Website www.poultryclub.org/minorcaclub
Tel 07767 237840 and 01885 410453

The Modern Game Club
Jennifer O'Sullivan
42 Sussex Avenue
Ashford
Kent TN24 8NB
Website www.moderngameclub.co.uk
Email moderngame.club@freeola.net
Tel 01303 813428

The New Hampshire Red Club
Mrs C. Compton
Devonia
Northbrook
Micheldever
Winchester
Hampshire SO21 3AH
Tel 01962 774476

The Old English Game Bantam Club
Mr M. Woolway
31 Pencefnarda Road
Penrheol
Gorseinon
West Glamorgan SA4 4FY
Tel 01792 894433

The Orpington Club
Andrew Richardson
Black Lane
Head Farm
Nately
Preston
Lancashire PR3 0LH
Email andrew.richardson@virgin.net
Tel 01253 790468

The Pekin Bantam Club
Mr D. Sill
Wards End Farm
Marsden
Huddersfield
West Yorkshire HD7 6NJ
Tel 01484 841008

The Rare Poultry Society
Mrs Anne Merriman
Danby
The Causeway
Congresbury
Bristol BS49 5DJ
Website www.rarepoultrysociety.co.uk
Tel 01934 833619

The Rhode Island Red Club
Norman Steer
Crossways
Kerries Road
South Brent
Devon TQ10 9DE
Tel 01364 732946

The Sebright Club
Steve Fuller
1 Ridge Farm Cottages
Rowhook
Horsham
West Sussex RH12 3QB
Tel 01306 628369

The Silkie Club of Great Britain
Miss Louise Hidden
67 Glossop Road
Charlesworth
Glossop
Derbyshire SK13 5HF
Website www.thesilkieclubofgreatbritain.co.uk
Tel 01457 855720

The Sussex Club
Miss Sarah Raisey
Rosebeam Farm
Burlscombe
Tiverton
Devon EX16 7JJ
Tel 01823 672789

The Plymouth Rock Club
Mrs Sally Prescott
St Helen's Cottage
Kickinsons Lane
North Thoresby
Grimsby
Lincolnshire DN36 5RG
Email prescott@thoresby.freeserve.co.uk
Tel 01472 840142 (evenings)

The Poland Club of Great Britain
Mrs Amelia Richardson
11 Eastfields
Martock
Somerset TA12 6NW
Tel 01935 827845

The Turkey Club UK
Mrs J. Houghton-Wallace
Cults Farmhouse
Whithorn
Newton Stewart
Dumfries and Galloway DG8 8HA
Website www.turkeyclub.org.uk
Email janwallace@aol.com
Tel 01988 600763

The Welsummer Club
Geoffrey Johnson
Aston Mews
Coppice Green Lane
Shifnal
Shropshire TF11 8TP
Website www.welsummerclub.org
Tel 01952 460274

The Wyandotte Club
Jeff Maddock
28 Lapstone Road
Millom
Cumbria LA18 4BU
Tel 01229 772556

Glossary

Bantams – miniature breeds of chicken, approximately one-quarter the size of their large fowl counterparts; 'true bantams' are breeds that have no large equivalent.

Beard – head feathers beneath the beak.

Beetle green – the iridescent green sheen exhibited by black feathers when struck by light.

BHWT – Battery Hen Welfare Trust.

Blood ring – a red ring visible through the shell of an egg when the embryo has died.

Bloom – protective layer deposited on an egg at laying.

Breeders' pellets – a prepared feed ration, suitable for birds being used for breeding.

Breeding pen – another term used to describe a breeding flock, usually consisting of one cock and several hens.

Broiler – a bird bred specifically for the meat trade.

Brooding – the sitting process by which a chicken incubates its eggs.

Broody – the stage in the reproductive cycle at which the hen incubates eggs.

Caeca – paired, blind-ended part of the gut where digestion of cellulose takes place.

Caecal dropping – every seventh to tenth dropping voided by a chicken comes from the caeca or blind gut; it's of a foamy consistency, easily distinguishable from normal droppings.

Candling – the process of checking the progress of a developing egg by holding it to a beam of light to illuminate its contents; also used to check for cracks.

Capon – a castrated cock fattened for meat.

Chick crumb – chicken feed with high protein content, broken up small enough to be eaten by chicks.

Chook – slang for a chicken, used mainly in Australia.

Cock – a male chicken over one year old, normally past its first annual moult.

Cockerel – a male chicken under one year old, bred in the current year.

Comb – fleshy head ornamentation, often larger in males.

Coverts – small covering feathers on the wings and tail.

Crest – a display of feathers on top of the head.

Cushion – the posterior of the back in the female; the male equivalent is the saddle.

DEFRA – Department of the Environment, Food and Rural Affairs.

Dewlap – a fold of fleshy skin under the chin.

Down feathers – fluffy feathers that have no shaft, typical of chicks.

Drinker – the container in which water is placed.

Dubbing – removing the comb and wattles of a cock.

Ear lobes – the flesh below the ear canal.

Egg tooth – the protuberance on a chick's beak used in 'pipping'.

Feeder – the container in which feed is placed.

Flight feathers – the first ten feathers of the wing, also termed the primaries.

Free-range – access to ground with a large percentage of vegetation.

Frizzle – feathers which curl backwards towards the head, giving the chicken a 'permed' appearance.

Fur and feather – traditional reference to rabbits and chickens.

Game birds – birds raised and shot for sport, including pheasant, partridge, grouse, quail and guineafowl.

Go broody – term used to describe a hen's desire to incubate her eggs.

Grower – a young chicken sold at the age of usually about six weeks.

Growers' pellets – prepared feed suitable for growing birds.

Gular flutter – term used to describe the fluttering of a chicken's throat that helps it to cool down.

Hackles – neck feathers.

Hard-feathered – close, tight feathering as in game birds.

Hen – a female chicken over one year old, normally past its first annual moult.

Henny-feathered – a male having rounded, hen-like feathers.

Hybrid – a bird with parents from two different breeds; used today as a term for commercial laying birds.

Inbreeding – breeding mother to son, father to daughter.

In lay – term used to describe a chicken that's laying eggs.

Keel – the blade-like ridge of a chicken's breastbone; in ducks and geese it also refers to the flap of skin hanging below.

Layer – an egg-producing chicken.

Layers' pellets – prepared feed suitable for birds laying eggs.

Line breeding – see inbreeding.

Litter – the dry medium applied to the floor of a coop, normally wood shavings, straw or short-shredded paper.

Mash – powdered chicken feed.

Miniature – alternative term for a bantam.

Moult – the process of annual feather replacement.

Muff – feathers on a chicken's cheeks, attached to the beard; also called whiskers.

Muffling – beard and whiskers.

Ornamental – a breed only suitable for exhibition.

Paunch – in ducks and geese, the pouch of skin that hangs between the legs.

PCGB – Poultry Club of Great Britain.

Pellets – chicken feed shaped into grain-sized pieces.

Pipping – the process by which a chick breaks into the air sac within its egg prior to hatching.

Point of lay (POL) – the stage of its life at which a chicken can start to lay eggs, usually at the age of about 16–20 weeks.

Pop-hole – the chickens' entrance into their house or run, usually closed by means of a door or sliding shutter.

Poult – a young turkey.

Poultry saddle – a cloth saddle placed on the back of a hen to protect her from the male's spurs and claws; can also be used on a rooster to protect it from rough treatment such as feather pecking.

Primaries – the first ten feathers of the wing, also termed the flight feathers.

Probiotic – beneficial gut bacteria, normally given in the drinking water.

Pullet – a female chicken under one year old, bred in the current year.

Saddle – the posterior of the back in the male; the female equivalent is the cushion.

Setting – the process whereby eggs are placed in an incubator to commence their development.

Sitter – term describing a bird used to sit on eggs, often applied to a duck.

Sitting – the process of settling on eggs to incubate them.

Soft-feathered – loose, fluffy feathering typical of Orpingtons.

Stag – a male turkey.

Table bird – a bird reared for meat.

Top-knot or tuft – other names for a chicken's crest.

Trap nest – the process whereby a hen is blocked into its nest box for monitoring etc.

Trug – a rubberised bucket.

Urates – a chicken's urine, solid and white.

Vulture hocks – stiff, downward-pointing feathers on the lower leg at the hock, deemed a fault in most birds, but part of the breed standard for the Sultan.

Wattles – the coloured fleshy lobes that hang from a chicken's chin either side of the beak; more pronounced in the male than the female.

Whiskers – feathers on a chicken's cheeks, attached to the beard; also called the muff.

Withdrawal period – the legally determined length of time between a bird being given its last dose of medicine, and its meat or eggs being deemed fit for human consumption.

Zoonotic – diseases that can be passed to and from humans.

Further reading

Avoid the Vet: Practical Poultry (Kelsey Publishing, 2008)

Ducks and Geese: A Guide to Management (Crowood Press, 1991) Bartlett, Tom

The Polish Breed of Poultry (Beech Publishing House, 2002) Batty, Joseph

The Silkie Fowl (Beech Publishing House, 2008) Batty, Joseph

Creative Poultry Breeding (Veronica Mayhew, 2005; available from veronica.mayhew@virgin.net) Carefoot, W.C.

The Chicken Health Handbook (Storey Publishing, 2008) Damerow, Gail

Extra Extraordinary Chickens (Harry Abrams, 2005) Green-Armytage, Stephen

Bantam Breeding and Genetics (Spur Publications, 1977) Jeffrey, Fred P.

Poultry House Construction (Gold Cockerel Books, 2005) Roberts, Michael

Diseases of Free-Range Poultry (Whittet Books, 2009) Roberts, Victoria

British Poultry Standards (Blackwell Science, in association with the Poultry Club of Great Britain, 2008) Roberts, Victoria

The New Duck Handbook (Barrons, 1989) Raethel, Heinz-Sigurd

Incubation: A Guide to Hatching and Rearing (Broad Leys Publishing, 1997) Thear, Katie

Starting with Chickens (Broad Leys Publishing, 1999) Thear, Katie

The Complete Encyclopedia of Chickens (Rebo Publishers, 2004) Verhoeff, Esther, and Rijs, Aad

Acknowledgements

KJanice Houghton-Wallace – Turkey Club UK (www.turkeyclub.org.uk)
Omlet UK (www.omlet.co.uk)
Verm-X (www.verm-x.com)
Karen Power – Power Poultry Norfolk (www.powerpoultry.co.uk)
Leigh Morris– Gothic Farm, Holbeach Drove
Pat Wilson – Weight Loss Resources, Peterborough (www.weightlossresources.co.uk)

USDA – www.usda.gov
Matt Mead – Windrush Poultry (www.windrushpoultry.co.uk)
Freddie Watts – Arthurs Place Boarding Cattery (www.arthursplacecattery.co.uk)
The members of the Practical Poultry forum (www.practicalpoultry.co.uk)
Manor Farm Poultry (www.manorfarmpoultry.co.uk)
Jenni O'Sullivan (www.moderngameclub.co.uk)
Phil Smedley, The Ancona Club, York